Who Killed
the Grand Banks?

Who Killed the Grand Banks?

The Untold Story Behind the Decimation of One of the World's Greatest Natural Resources

Alex Rose

WILEY

John Wiley & Sons Canada, Ltd.

Library and Archives Canada Cataloguing in Publication Data

Rose, Alex
   Who killed the Grand Banks? : the untold story behind the decimation of one of the world's greatest natural resources / Alex Rose.

Includes bibliographical references and index.
ISBN 978-0-470-15387-1

   1. Atlantic cod fisheries–Closures–Atlantic Coast (Canada). 2. Grand Banks of Newfoundland. 3. Fishery management–Canada. 4. Fishery policy–Canada. 5. Natural resources–Government policy–Canada. 6. Natural resources–Canada. I. Title.

SH229.R58 2008   333.95'66330916344   C2008-901479-0

Production Credits
Cover: Adrian So
Interior text design: Tegan Wallace
Typesetter: Thomson Digital
Printer: Friesens

John Wiley & Sons Canada, Ltd.
6045 Freemont Blvd.
Mississauga, Ontario
L5R 4J3

This book is printed with biodegradable vegetable-based inks. Text pages are printed on 55lb 100% PCW Hi-Bulk Natural by Friesens Corp., an FSC certified printer.

Printed in Canada

1 2 3 4 5 FP 12 11 10 09 08

To Joanne, Caroline, and Alexandra

Table of Contents

Acknowledgements

When it comes to the near extinction of the Grand Banks cod—for 500 years one of the world's known natural wonders—people in Newfoundland and governments in St. John's and Ottawa set up an impenetrable defence. Perhaps shamed and embarrassed by their complicity in a biological disaster to rival anything in recent history, many have taken refuge in denial, obfuscation and a self-righteousness that always blames the other; the naturally garrulous people of Newfoundland have been reduced to a stony silence. This made research maddening, circular, Sisyphean. But others, undaunted by academic careerism, pensions, pay cheques or party lines, worked overtime to help me uncover the facts. Two seminal books—*Fishing for Truth* by Alan Christopher Finlayson and *Useless Arithmetic* by Orrin H. Pilkey and Linda Pilkey-Jarvis—helped frame my initial research. From the *Illustrated History of Canada*, I am especially indebted to historians Arthur Ray and Graeme Wynn. Later, Daniel Pauly and Rashid Sumalia of the Fisheries Centre at the University of British Columbia patiently shared new research with me as did their colleagues Carl Walters and Sylvie Guénette. Through a different lens, resource economist Peter Pearse argued that quotas will, in the future, help to prevent other fisheries from "going the way of the Grand Banks cod."

Two people helped me understand the link between the Grand Banks cod slaughter and issues such as East Coast offshore oil and gas activity (Moira Baird, Newfoundland) and fast-declining Pacific salmon stocks (Scott Simpson, *Vancouver Sun*, British Columbia). Stewart Bell, Helena Bryan, Johanne Fischer, Peter Hill, Bonnie Irving, Bob Laughlin, Glen Power, David Stouck, and Phil Wallace provided advice and encouragement throughout and I would not have been able to finish this book without the generosity and patience of Kim Mah. I have benefitted from them all and it is a pleasure to list them here, while exonerating them from the errors and omissions that, like submerged fishnets, may survive as the book goes into production.

1

A Great English Ship Moored Near the Grand Banks

"The cod were so thick we hardly have been able to row a boat through them."
— John Cabot

Head northeast in Newfoundland along the Bonavista Peninsula, until the rough road ends and the cold ocean begins, and you'll find a bronze statue of a man clothed in puffy Renaissance garb overlooking tumultuous seas. The landscape has a distinctly untamed feel to it as icebergs float offshore like errant mountaintops, humpback whales feed and breach, and puffins dart along the ocean surface to and from their island colony just around the point.

That improbably dapper man is John Cabot who, in 1497, reached the New World—the New Founde Land, as it was dubbed. Like Columbus five years before him, Cabot was an Italian sailor seeking a shortcut to the Orient. He figured that by heading north to where the longitudinal lines were closer, he'd shave off some sailing days on his westward voyage to Asia. He figured wrong, of course. Cabot and his crew of eighteen ran into the invincible coast of eastern Canada and, instead of spices and porcelain, they found spruce, the oldest granite on Earth, and ice floes. Undaunted, Cabot claimed the region for the English throne, which had financed his expedition, then headed home.

So enduring are the tales of Cabot's arrival hereabouts—the peninsula's name is derived from the first words he was said to have uttered—that it seems unfair to mention that Cabot's landing spot is wholly a matter of conjecture. Historians say Cabot may very well have made landfall around the Bonavista Peninsula, or perhaps in Nova Scotia on Cape Breton Island, where a similar statue also marks the event. Then again, maybe Cabot landed in Labrador (now part of the Province of Newfoundland and Labrador), or much farther south in Maine. No one knows for sure.

While Cabot's landfall may be in dispute, what he discovered is not: cod—and lots of them.

Every school child knows the story. Five hundred years ago the explorer John Cabot returned from the waters around present-day Newfoundland to report that the codfish ran so thick they were easily caught by dangling a wicker basket over the side of the vessel. The log of Cabot's ship, the *Matthew*, reported there were six- and seven-foot-long codfish weighing as much as 200 pounds. Cabot discovered a resource that would shape world politics for hundreds of years, launch a fiercely competitive Maritime trade, and create safe harbours along the shores of the new colony of Newfoundland: the limitless bounty of the Grand Banks cod.

Cabot led two explorations from Bristol, in 1497 and 1498. King Henry VII, who had agreed to his voyage and helped to pay for it, rewarded Cabot with the sum of £10. On his return to England, Cabot related amazing tales of this new world. Like witnesses in the New Testament, Cabot told tales of men who could walk across the Grand Banks waters on the backs of cod. Fictional, perhaps, and more likely sales patter for his clients, this fishy narrative was spun into a tale of mythic proportion, painting a false reality of this foreboding, rock-ribbed place moored in the North Atlantic.

Historic accounts say that Cabot lowered a basket weighted with stones into the North Atlantic, then hauled it back up brimming with cod. The discovery of these fertile fishing grounds set off a centuries-long struggle among Basque, Portuguese, French and English fishermen, and established a pattern of far-flung coastal settlements, called outports by Newfoundlanders, that ring the island.

And so the legend fits today: the Grand Banks became Valhalla, a miraculous, self-sustaining Eighth Wonder of the world, feeding the known world for 500 years.

The catastrophic collapse of the fisheries, circa 1992, was unprecedented. An ecological disaster to rival any other—the destruction of the Amazonian rain forest notwithstanding—in modern history. This made-in-Canada plunder was part human greed, part stupidity, and part rapacity. Tarnishing Canada's standing within the international community, it holds the reputation of Canada's once-vaunted fisheries scientists up to ridicule. Sixteen years later, no one has taken accountability or apologized for the ruination of a centuries-old way of life and, more shocking, a stock recovery plan has yet to be produced.

The plunder was born of a mindset that has much to do with survival. And that mindset still has currency today in a saying often heard in the coffee shops and bars of this rock-ribbed island: "If it runs, walks, or swims, kill it." The Grand Banks cod was the *raison d'être* for the economic lifeblood and culture of Newfoundland. Little wonder that governments in St. John's and Ottawa were ever-devising and fine-tuning policies that worked overtime to build an economy from this one resource. Today, many Newfoundlanders, resorting to the blame game, no longer want to talk about the destruction of the ecosystem. A naturally talkative community has been reduced to a stony silence.

Visitors wonder if Newfoundlanders share a sense of collective shame as they resort to pat and automatic answers about what happened to the cod. Fingers are pointed at foreign fishing, seals, changes to water temperature, botched science, bad management by federal fisheries, and, as always, the politicians in Ottawa. Today, the bereft citizens just want to move on, determined to allocate the Grand Banks cod collapse to memory.

There can be no forgetting—or forgiving—such catastrophic pillaging. Sparked by a second wave of environmentalism focusing on the state of the world's oceans, the Grand Banks cod collapse became a talking point, a *sujet noir,* now studied at universities and fisheries research centres, wherein students from around the world routinely repeat: we must never allow our fisheries to go the way of the Grand Banks cod—once the largest and most productive fishery in the world.

Fished by European nations since the 1400s—and joined by the Russians, Japanese and Koreans in modern times—the vastness of the cod supply led many to believe that it was an inexhaustible resource. Indeed, Professor Jeffrey Hutchings of Dalhousie University once estimated the theoretical catch since the 1400s might weigh in the billions of tonnes. The unprecedented richness of the cod stocks galvanized repeated military conflicts between nations for the control of access to these fishing grounds, the Canadian federal government to establish off-shore limits, and European colonization of an otherwise barren and inhospitable land.

It is difficult to over-emphasize the importance of the fishery to the people of Newfoundland, which became a province of Canada in 1949. Since Newfoundland's confederation with Canada, the cod fishery remains the single most powerful source of collective cultural identity for the people born and raised there.

The story begins with England's 300-year exploitation of Newfoundland, and provides insight into widely held resentments that play out today: mistrust and suspicion of all things Ottawa, a legacy of the cruel colonial master England once was. Resentments that are grudgingly accommodated as long as federal dollars keep flowing.

England's presence in the Newfoundland fishery increased steadily from the mid-16th century, while the Portuguese and Spanish fisheries waned towards the century's end. The French and English wrestled for control over the fishery throughout the 17th century. After the exclusion of the Spanish and Portuguese, the English laid claim to the shore from Cape Bonavista to Cape Race. Between 1610 and 1623, land grants issued by the English-owned Newfoundland Company resulted in the establishment of permanent and semi-permanent residences. By the 1670s, there were as many as seventeen English communities along the coast from Trepassey to Bonavista.

Through the 16th and most of the 17th centuries, fishing remained mostly a transient industry, despite some attempts to promote settlement. In 1610, John Guy of Bristol led a party of settlers to Conception Bay, and in the next decade Lord Baltimore began a short-lived settlement of English Catholics at Ferryland on the Avalon Peninsula. The French established stations on the south coast along the shores of Placentia and Fortune Bays, land that was en route to their settlements along the St. Lawrence River in New France. French Basques were the primary fishermen on the south coast, from Cape Race to Lamaline. While the Normands and Rochelais were engaged largely in the bank fishery, fishermen based in and around St. Malo continued to fish along the Petit Nord and along a few locations on the Strait of Belle Isle south to Cape Ray.

The English fishery expanded rapidly during the late 16th and early 17th century, but declined after 1624, partly because of wars with France and Spain and partly because of increased activity of pirates. An important consequence of these events was the slow, but steady rise of year-round resident fishermen known as boat-keepers or planters. As a result, catches by planters rose from zero to 37 per cent more than the English fishing ships between 1610 and 1675. By the late 1600s, English Newfoundland comprised the eastern shore of the island, from Trinity Bay and Conception Bay to Ferryland and Renews, south of St. John's.

Here, on what is today's Avalon Peninsula, barely a thousand men, with a few women and children, might winter every year. They were joined every summer by thousands of fishermen from England. St. John's, a rendezvous for fishermen since the 1500s, was already the largest settlement, but people were scattered among a score of outports, wherever there was a harbour and handy access to the fish. French fishermen, travelling back and forth from Basque, Breton, and Norman ports, fished each summer on the northern coast of Newfoundland. In approximately 1660, a small French settlement called Plaisance, complete with governor, garrison, fortifications, and a few hundred people, was constructed on the south coast of the island.

Traditionally regarded by British officials as "a great English ship moored near the Banks" for the convenience of the fishery, cold, inhospitable Newfoundland was settled slowly. In Britain, the migratory fishing industry was considered a vital nursery of seamen, turning "green men" into "salty dogs" who could man the navy in times of crisis. It was also a lucrative trade on which the fortunes of many an English merchant was built. In the early 1760s, the number of people who wintered in Newfoundland rose to between 8,000 and 9,000. Permanent residents lived where the fishing was best. Scattered along the shore were small, strand-clinging clusters of dwellings, sheds, stages (or wharves), and flakes (or drying racks) that tied sea to land and became the focus of summer work.

From each of these villages, day after arduous day (May through September) small boats with crews of three, sometimes four, were rowed out to local fishing grounds and filled with cod taken on baited lines. Boats from English vessels, moored inshore and unrigged for the summer, worked alongside those of resident fishermen. Each evening the catch was split and salted; each morning it was spread and turned on the flakes to dry, until, in late summer or fall, it was shipped to market. Neither rhythm nor routine varied much. Here, from outports all along the Avalon Peninsula, fifty boats might set out, there barely a dozen. But skills and circumstances were remarkably uniform.

By the 1770s, the fishery comprised three distinct groups. Merchants, usually resident in England or Ireland, organized the trade, operated stores in Newfoundland, and participated in a far-flung network of international exchange. Boat keepers (or planters), either resident or migratory, owned the boats and equipment of the inshore fishery, and acquired supplies from the merchants to whom they sold their catch. Servants (resident or migratory) made up the third group. They fished for the boat keepers or those merchants who operated boats on their own account. But the shrinking pool of indentured or migratory labour, and rising equipment and food costs (which doubled while fish prices increased by only half), squeezed the boat keepers especially hard.

More frequently, they drew their crews from among their kin, while wives and children tended the catch on shore. In effect, the boat keepers slid down the status ladder to become ordinary fishermen. As they did so, they fed themselves by cultivating potatoes and garden crops, keeping a sow or two, hunting, and perhaps gathering wild fruits and berries. Coincidentally, seal fishing increased and the returns usually provided small, but valuable supplements to the fisherman's earnings.

Settlement on the island, which seemed to threaten both the security and the profit of the English, was discouraged. Between 1760 and 1780, the Newfoundland fishery was essentially a seasonal enterprise conducted from Europe. During the American Revolution, the fishery was harassed by press gangs and privateers, and with the resumption of war between Britain and France in 1793, the number of people in this migratory fishery declined.

The disapproval of traders and politicians could not prevent a steady growth of population during the latter half of the 18th century. By 1800, permanent residents made up 90 per cent of Newfoundland's summer population and produced 95 per cent of its codfish exports. Suddenly, reported a British naval officer, the island had "more the appearance of a Colony than a fishery from the great number of People who have annually imperceptibly remained the Winter and who have Houses, Land and Family's," a transformation quickly consolidated by a dramatic increase in the proportion of women and children migrating there.

When the fishing fleets left in the fall, men who had been contracted as ship's crew for two or three summers wintered in Newfoundland to guard and maintain the shore-works required to cure and keep the summer's catch. Some found wives among the female domestics brought to the island by military officers and other officials, and sought a permanent life there. Others preferred Newfoundland to the prospect of unemployment in Devon or starvation in Ireland.

St. John's, with a British military garrison of approximately 200 and a cluster of merchants' stores, became the commercial centre of Newfoundland. There, as in the smallest settlements, cattle, sheep, and fowl wandered the rough paths that linked buildings, and foraged in the brush beyond. "For dirt and filth of all kinds," wrote Joseph Banks, the botanist in Captain Cook's expedition to Newfoundland in the 1760s, "St. John's in my opinion may reign unequalled."

Settlements along the coasts, the outports, received supplies by sea from St. John's and forwarded their catch directly to the merchants there. Barter replaced cash in these exchanges. Gradually the number of outport merchants fell and the artisans who had built the boats and made the barrels for the fishery largely disappeared from the outlying settlements, as families themselves took up these crafts in their increasingly closed, self-sufficient communities. The scattered, isolated settlements of Newfoundland began to assume their typical 19th- and 20th-century character. Life within these communities, dependent on the resources of a narrow band of land and sea, and centred around families closely linked by blood and marriage, was intensely local and highly traditional. Each family turned its energies to an enormous variety of tasks during the year and each member of the household developed a wide range of skills, from mending nets and shearing sheep to curing fish and barreling pork.

From such a demanding routine people gained a living but little more. Dwellings were modest. Many, said a visitor to Trinity Bay in 1819, "consist only of a ground floor"; although the best of them were clapboarded, most were "built of logs left rough and uneven on the inside and the outside," and they had "only one fireplace in a very large kitchen." Material comforts were few and any decline in the price or the catch of fish was likely to produce "appalling poverty and misery."

In the mid-1800s, fishing remained dominated by fleets from England, France, Spain, or Portugal, which arrived each spring, and returned home in the autumn with their dried or brine-soaked catch. Cod was at least as expensive as beef, and Europe lacked the means to transport it far inland, so only a minority of European people ever ate Newfoundland cod, yet the cod trade was always much larger and more valuable than the fur trade, historians estimate. Along the English Channel, the preference was for "green" cod lightly pickled in brine, but most Newfoundland cod was split, salted, and exposed to sun and air until it was thoroughly hard and dry. Dried cod could be preserved for months or years, and it was this dried product from North America that opened a market for cod on the hot southern coasts of Europe. French and English fishermen competed for this market in Portugal, Spain, and into the Mediterranean.

Drying stations were established on the Newfoundland coast, particularly on the Avalon Peninsula, at good harbour sites that were also favourite camping places of the Beothuk. In the early 16th century, fishermen had little interest in trading with them. The Beothuk did not welcome the fishermen because they occupied Native campsites and destroyed the surrounding forests with their clearing and reckless burning. Conversely, the fishermen disliked the fact that during the winter the Beothuk frequently plundered the drying stations to obtain nails and other metal scraps.

This small tribe suffered in ensuing hostilities and, by the early 19th century, became one of the few aboriginal groups in Canada to be totally annihilated. Today, historians and First Nations have squared off in a heated debate as to whether the Beothuk were victims of genocide.

The Newfoundland fishery for Atlantic cod remained the largest and the most productive cod fishery in the world throughout the great age of the Grand Banks schooner and well into the 20th century. But, by the second half of the century, technological advances in fishing equipment such as radar, depth finders, and sonar built into the bridge of massive new trawlers and factory shops, as well as a frenzy of competition with other fishing nations, led to the most volatile period in the Newfoundland fishery. The Grand Banks were plundered with a selfish greed unseen until the early 1990s, after which the cumulative biological effects of three decades of overfishing resulted in the imposition of a commercial moratorium on July 2, 1992.

Northern cod had been fished since the late 15th century, but total harvests appear to have been less than 100,000 tonnes annually until the late 18th century. In the 1880s and 1910s catches increased to as much as 300,000 tonnes, before declining to less than 150,000 tonnes in the mid-1940s.

Even as early as the First World War, trawler vessels towing large conical nets along the sea bottom were becoming significant in the groundfish industry. In the 1920s, Newfoundland fisheries were served mainly by old fishing craft from Nova Scotia while European fleets used the more reliable trawlers. Following the expansion of European-based factory trawlers in the late 1950s and early 1960s—the single most important event in the 500-year history of the Northern cod fishery—catches increased dramatically to historic maximums in 1968. Thereafter, the catches collapsed in equally dramatic fashion until 1977 when Canada extended its fisheries jurisdiction to 200 nautical miles.

The first of these enormous vessels (up to 8,000 Gross Registered Tons (GRT)), the British ship *Fairtry* appeared in 1954 and was soon followed by two Soviet trawlers in 1956 and a West German trawler in 1957. The unprecedented harvesting capacity of these vessels became evident in the 1960s when both catches and the number of fishing nations reached levels never before attained. Between the mid-1950s and late 1960s, the total catch of Northern cod by all countries almost tripled, reaching 810,000 tonnes in 1968. Catches plummeted thereafter to 139,000 tonnes in 1978. Controlled in part by the Total Allowable Catches (TACs) established by the Canadian government (the inshore component of the Northern cod fishery has never been limited by catch quotas), catches increased gradually to a post-1977 high of 268,000 tonnes in 1988, prior to the imposition of the moratorium.

All the above catch estimates are almost certainly underestimates due to discarding of unwanted fish and misreporting. The dramatic changes that occurred in the Northern cod fishery were precipitated by trends that began in 1959. For probably the first time in the history of the fishery, offshore catches exceeded inshore catches and, for the first time since the early 1700s, catches by foreign fleets exceeded those of resident Newfoundlanders.

Twenty different countries prosecuted the Northern cod fishery between 1954 and 1990. In 1954, Northern cod was fished by Canada, France, Portugal, Spain, and the United Kingdom. By 1959, these countries were joined by the USSR, East Germany, West Germany, Norway, and Iceland. With the exclusion of Norway and the inclusion of Poland and Japan, these were the eleven nations fishing for Northern cod when the historic high of 810,000 tonnes was taken in 1968. The greatest number of countries participating in the Northern cod fishery in any single year occurred in 1977 when sixteen countries were involved (oddly the first year of Canada's jurisdictional extension to 200 miles). With the exception of Iceland and the addition of Norway, Cuba, Italy, Romania, Ireland, and the Faroe Islands, the nations fishing in 1977 were the same ones doing so in 1968. All this has defined and roiled Newfoundland politics to this day.

Several hundred years before confederation with Canada, Newfoundland was a functioning European settler society with a distinct language and culture, clearly defined borders, laws, and institutions. The capital, St. John's, was founded in 1583 and is the oldest city in North America. Its first civil court was established in 1791, and its first chief justice was appointed in 1792. A system of governors was established in 1729, the first popularly elected legislative assembly convened in 1832, and responsible government was achieved in 1855.

From that date, Newfoundland was essentially a country and, except for the power to sign international treaties, it exercised all the normal powers of sovereignty that Canada did, including self-defence and the issuing of postage and legal tender. There's even a national anthem—the beloved "Ode to Newfoundland"—that is regularly performed at official ceremonies and still moves some to tears.

In 1949, spearheaded by a fast-talking, pig-farmer-turned-politician-crusader named Joey Smallwood, the province voted to join Canada, but only by a narrow 4 per cent difference. In his own words, Smallwood was determined to drag Newfoundland "kicking and screaming into the twentieth century." He was responsible for shutting down several hundred remote outports, forcing residents to relocate to "growth centres" such as Trepassey. In December, 1991 he died at his home outside St. John's at the age of ninety.

Smallwood, known as Joey, was a salty figure who, during twenty-two years as Newfoundland's premier, both infuriated and delighted Canadians with his efforts to extend the province's influence across the country. Newfoundland, which with Labrador was a British colony, decided to join the Canadian confederation in a 1948 referendum by a narrow 7,000-vote margin. Smallwood led the campaign for joining Canada against opponents who favoured an economic union with the United States or that Newfoundland declare its independence.

Not one to offer smooth comforts when blunt words would do, Smallwood made his mark by telling voters before the referendum that a rugged self-assurance born of centuries of hard-scrabble existence, mostly from fishing the rough seas of the Grand Banks, did not qualify them for independence. "We are not a nation," he told the Newfoundlanders, many of whom were of Scottish and Irish descent. "We are a medium-sized municipality left far behind the march of time." Asked years later what would have become of Newfoundland had it not joined Canada, Smallwood replied, "We'd be bankrupt and destitute."

The referendum ended nearly 400 years of British rule over the island of Newfoundland and the adjacent mainland territory of Labrador, which together became Canada's tenth and—at the time—most have-not province. This is a term to measure the economic needs of a province within Confederation, triggering funds from the federal government. This was Newfoundland's lament until recently when the province began to tap into its rich offshore oil and gas deposits. Seemingly overnight, it became a "have" province with new revenues offsetting the calamitous economic impacts resulting from the collapse of the Grand Banks cod.

In 1941, a new Unemployment Insurance Act, later renamed Employment Insurance in 1996, was introduced to Canada. This system is an important component of the economic safety net provided by government and there is little disagreement, in principle, that it has provided greater income security for Canadians. Among economists, however, there is substantial concern that specific features of the existing system may create unemployment. For example, it has been argued that unemployment is higher than it should be among those employed in seasonal industries because it may be easier to collect benefits than to look for other work during the off-season. In the case of Newfoundland, many historians believe that the fishers became less self-reliant over time and began to see state-supported assistance as a routine supplement to fluctuating incomes from the fishery; far too many Newfoundlanders are aggressively dependent on federal government hand-outs which, for many in the outports, became a way of life. Unfortunately, there is little data to confirm that the abuse of unemployment insurance benefits is more or less than anywhere else in Canada.

In 1977, Canada extended its jurisdiction to 200 miles creating a euphoric, gold-rush mentality, sanctioned and torqued by policies in Ottawa and St. John's. The Grand Banks stocks were relentlessly hammered into oblivion for the next fifteen years. Canadians continued catching cod at absurdly high and unsustainable levels long after the foreigners were kicked out.

On July 2, 1992, a reluctant Federal Fisheries Minister, John Crosbie, announced the closure of the Northern cod fishery in St. John's, Newfoundland. With this announcement, he as much as admitted that the fisheries management scheme set up by Ottawa in 1977, after the declaration of the 200-mile fishing zone, had failed. Of course, Mr. Crosbie did not admit that government policy was at fault. Instead, he continued to advance the official government line that the collapse of the fishery was attributed to three causes: foreign overfishing outside of our 200-mile limit, a rapid increase in the seal population, and oceanographic phenomena such as cold water barriers. All three causes were indisputably suspect.

Today, despite a new euphoria—boosterism along Water Street in St. John's is hoisted like a huzzah with every pint—the province still has the nation's highest unemployment rate, lowest per-capita income, some of the highest rates of taxation, highest per-capita debt, weakest financial position, highest rate of out-migration, and fastest population decline. Since the moratorium, 70,000 Newfoundlanders, or 12 percent of the population, have "outmigrated" to the oilfields of northern Alberta and the factories and retail counters in Ontario's industrial triangle. A $700-million industry evaporated overnight.

The decline was greatest after the cod moratorium, but it persists today. And it will continue. Today's population is about 510,000. Under the best-case scenario, it will fall to 495,000 by 2020; under the worst, 460,000. In a generation, Newfoundland has gone from having the country's youngest population profile to its oldest. In the 1970s, Newfoundland had the country's highest birth rate. It now has the lowest, due to the steady stream of people of child-rearing age leaving the province.

This was hardly the expectation of the people of Newfoundland and Labrador when they entered Confederation in 1949. They surely imagined they would be forever fishing cod on the Grand Banks. But most of the cod fishery is closed; the dories sit unused on the beach while the offshore fleet lists idly at the dockside. Stocks are just 1 per cent of their numbers from 1977.

In 2007, a visitor from Canada's West Coast travelled to Newfoundland to ferret out the real story and the real people behind the official account of one of the worst ecological disasters in modern history. In the coffee shops of St. John's and in tiny outports that border the Grand Banks, he searched for answers to a mystery that has confounded people across Canada and around the world. It is important to solve a mystery that has shamed Canada's fisheries scientists, reduced an entire province to silence, cost Canadian taxpayers $6 billion, put much of an island on welfare, and destroyed a way of life forever.

Soon after his arrival the visitor learned that certain subjects enrage Newfoundlanders as they continue to reflect darkly on a commonly-held 500-year narrative, passed down from generation to generation, of exploitation by powerful outsiders.

Meanwhile, the Grand Banks are still being bulldozed. Newfoundland fisheries are more profitable now than they've ever been. The shrimp fishery is worth hundreds of millions of dollars. The near extinction of the cod made room for lower-tropic-level species that have since exploded in abundance, moving into the habitat where the cod once spawned, reared, and schooled. Shrimp and crab are more profitable than cod ever were. Shrimp, caught in high-tech Canadian trawlers, had a landed value of almost $400 million in 2006—about double the value of cod prior to the 1992 moratorium. Shrimp are loaded onto factory freezers and shipped to China, where they are hand-picked, then re-shipped to markets in the global marketplace.

Scientists have been warning for years that overfishing is degrading the health of the oceans and destroying the fish species—Canada's Grand Banks cod has become an international symbol of greed and stupidity—on which much of humanity depends for jobs and food, but still, the slaughter continues. And circumstances eerily similar to the plunder of the Grand Banks are being played out in Canada's Pacific salmon fishery, as an increasing number of experts warn that the salmon are "going the way of the Grand Banks cod."

2

The Cod: A Short History

"The docks reeked of the hybrid smell of fish and brine."
— Wayne Johnston

More than five hundred years ago, fishers from Portugal and the Basque region of Spain began fishing the fabled Grand Banks of Canada. Although many species of fish were harvested from the seemingly inexhaustible stock, the most famous and valuable was the cod. Thousands of vessels sailed back to Spain and Portugal, from the New World to the Old, their holds jammed with barrels of salted cod. Codfish—*bacalao* in Spain and *bacalhau* in Portugal—became a food staple for the entire Iberian Peninsula. Salted cod achieved added importance because of the numerous meatless days imposed by the Catholic Church. Later, generations of North American children learned of the importance of another cod product, the foul-tasting cod liver oil valued (by parents) as a source of vitamin D.

One 17th-century discourse on Newfoundland reported cod so dense that "We heardlie have been able to row a boate through them." All assumed that the cod stocks were boundless. And indeed, given the technological limitations of the fleets, they probably were.

An 1883 report on fisheries for the British Government asserted "The cod fishery . . . and probably all of the great sea fisheries are inexhaustible. Nothing we can do can seriously affect the numbers of fish." Two years later, the Canadian Ministry of Agriculture predicted, "Unless the order of nature is overthrown, for centuries to come our fisheries will continue to be fertile."

The Atlantic cod, *Gadus morhua* (Figure 2.1), has always been the mainstay of the Grand Banks fishery. Perhaps 90 per cent of the fish catch on the banks during the 1980s was cod. It is a tasty fish that can be salted or sun-dried and preserved for a long time, which was of particular importance in the days before refrigeration. Cod is often the fish used for fish-and-chips and for the McDonald's fish sandwich.

Cod were once found in schools, sometimes miles across, in deep water in the winter and in shallower water in the summer. The Atlantic cod probably has a number of subpopulations, each following the same migration paths year after year. The Northern cod used to extend from off the tip of Labrador down to Cape Hatteras off the Carolinas.

Huge agglomerations occurred during spawning and seasonal shoreward movements into shallower water during late spring, when the cod pursued capelin, a small, herring-like fish.

Several cod stocks have been identified throughout the vast range of the Grand Banks. While the species often moves upward in the water column in pursuit of food, cod generally prefer cooler water temperatures of 0° to 10° C, which is most often found at or near the ocean bottom in depths from 150 to 400 metres.

Figure 2.1: Atlantic cod

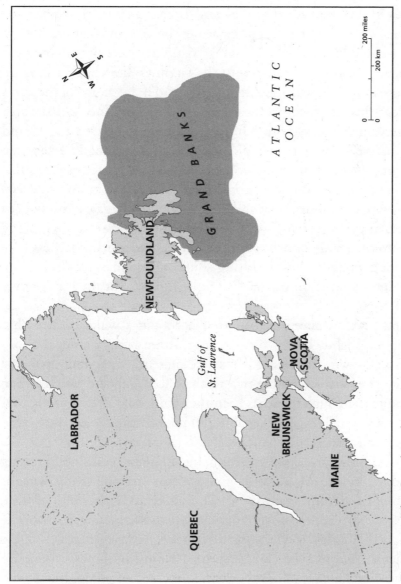

Figure 2.2: The Grand Banks

The Grand Banks (Figure 2.2) are located on the Canadian continental shelf off of Newfoundland. Nearly 300 miles across, it is one of the widest continental shelves of the world. The banks cover an area of 110,000 square miles and consist of shallow submarine plateaus, 75 to 300 feet deep, separated by troughs that are 600 or more feet deep. The cold Labrador Current flows down from the north, to mix over the banks with the warm Gulf Stream coming up from the south, causing the area to be draped in fog much of the time. The resulting churned-up waters are rich in nutrients and support a huge marine ecosystem. The conditions, unique to the Grand Banks, have created a splendid interconnected aquatic ecosystem that, for 500 years, fed much of the known world. Fish species like cod, haddock, and capelin once thrived, as did the bottom-feeders, the scallop and lobster. The area also supported large colonies of sea birds such as Northern Gannets, shearwaters, and sea ducks, along with various sea mammals: seals, dolphins, and whales. Icebergs are commonly present, slowly drifting south, losing mass along the way. The winter storms on the banks are legendary, but the water never freezes over.

Cod have an olive green spotted back and a white belly, with a prominent, slightly curved back-to-front stripe along the side. Various shades of brown and even red may be present, depending upon the habitat. They are commonly two to three feet long and weigh five to ten pounds, although there has been the occasional fish caught that was "as big as a man," six feet long and 200 pounds. And, cod continue to grow during their entire lifetime.

The cod eats just about anything, including the odd seabird resting on the rolling ocean surface. It is a fish that virtually swims with its mouth open, devouring clams, squid, mussels, echinoderms, jellyfish, sea squirts, worms, and other fish, including its own young. Its favourite fish is, perhaps, the capelin, a small plankton feeder that spawns in the summer on and near beaches. Capelin are most likely responsible for the cod's migration to shallow water in the summertime. Many who have written about the demise of the Atlantic cod have noted the irony that a fish as greedy as the cod is being destroyed by humans, another of God's creatures with even greater greed.

Cod spawn between March and June, releasing eggs that float to the surface and become part of the plankton for ten weeks. When the larvae reach one inch in length, they swim back to the bottom. Each female cod releases between 2 million and 11 million eggs—a stupendous figure that gave rise to a poem (said to be written by an anonymous American) comparing the productivity of codfish and chickens.

While reproduction peaks between February and April, cod are known to spawn over a much more extended period. Following fertilization, the eggs float to the surface, where they hatch when the embryos reach 3.3 to 5.7 millimetres in length. The larvae remain in the upper waters until they are between 25 and 50 millimetres long, when they drop to the bottom.

Growth rates vary with the stock and latitude. As a result of differences in water temperatures, the farther south the fish go, the faster their growth will be. Hence, a four-year-old fish in the northern part of its range might be 38 centimetres long, whereas a four-year-old farther south might be as long as seventy-one centimetres, almost double the length of its northern counterpart.

When harvest levels are low, Atlantic cod have been known to live as long as twenty-six years and weigh upwards of eighty kilograms. A rare sight today, however. Generally, commercially caught cod are five to six years old and weigh between 1.5 and 2.3 kilograms.

For centuries, Newfoundland fishermen, long lining from fourteen- and eighteen-foot-long dories, braved the frigid waves of the Grand Banks to earn a living. The boats were lowered from a mother ship each morning and gathered back in by nightfall. No other occupation could so exactly define such courage: to head out before daybreak in a bank of fog to snag the cod, pulling them goggle-eyed into the air, gills intoxicated with oxygen, then hauling them in, their bodies spanking the gunnels of the dories.

These tiny wooden vessels pitched in the rolling swells that could, in any instant, rise to enormous heights in violent gales. It was dangerous work immortalized by Winslow Homer's famous painting *Lost on the*

Grand Banks. The seascape shows two forlorn fishers, separated from the mother ship, peering over the side of their dory in rough weather. (Microsoft mogul Bill Gates purchased the painting in 1998 for $30 million. It was, by a factor of three, the highest price ever paid for an American painting.)

The story of this bravery is also told in the vivid, minor key evocations of songwriters Stan Rogers and Great Big Sea. Semi-fictional depictions of fishermen working on the Grand Banks can be found in Sebastian Junger's novel *The Perfect Storm,* in Rudyard Kipling's novel *Captains Courageous,* and in the achingly beautiful dreamscapes of Wayne Johnston's *The Colony of Unrequited Dreams*:

> The docks reeked of the hybrid smell of fish and brine. The wind off the cod was so salty that it made me sneeze and my eyes run with water. Sometimes when I breathed the air, my throat turned to fire, and when I tried to speak, I had no voice, could get out only a rasping, rattling cough as if I had inhaled a chest full of smoke. I never did acquire the ability to tolerate the omnipresent smell of cod. Walking among it always made me long for the clean-aired elevation of the Brow.

Gradually, newer and more efficient fishing methods came along. These include nearshore traps, used when cod come in to shallow water during the summer. Seines, which are nets pulled into circular traps by small motor vessels, and untended drift nets, are also both used on the Grand Banks. In some fisheries (not cod), drift nets can be as long as forty miles.

This method of cod fishing has been a particularly insidious and wasteful killer of Grand Banks fish. When the nets are lost or untended, large numbers of fish are caught by their gills as the net eventually sinks to the seafloor. Scavengers empty the net, which once again floats to the surface, fills with fish, sinks, and again returns to the surface after being emptied. The deadly cycle continues until the net disintegrates, which may take years if it is made of nylon.

Then came dragger technology, defined by the use of a powerful boat towing a large net bag along the ocean floor. The giant boards and rollers that keep the mouth open churn up the seafloor and destroy the plants and small animals essential for healthy marine life. Towed through dense masses of fish the dragger net acts as a non-selective vacuum cleaner sucking up big and small fish indiscriminately. Equipped with electronic fish-finding technology, the modern dragger is very good at locating and capturing fish, no matter where they are.

In fact, for the first half of the century, dragger technology was recognized by the Canadian fishing community as an ecologically unsound and socially destructive technology. Its use was prohibited by law.

The Grand Banks have historically been one of the world's richest fishing grounds. Portuguese and Basque fishers worked the Grand Banks as early as the 1400s. The area has since been fished by fleets from England, France, Spain, and Portugal, and later Newfoundland, Canada, Russia, and the United States, among many others.

When Canada followed the international trend and declared an offshore Exclusive Economic Zone of 200 nautical miles (370.4 kilometres) in 1977, about one-third of the fishable area of the Atlantic Continental Shelf fell outside this expanded jurisdiction. Included in this area are the Nose and the Tail of the Grand Banks, and the Flemish Cap, another fish-rich region east of the Nose. Non-Canadian vessels therefore fish the Nose and Tail and the Flemish Cap. The Northwest Atlantic Fisheries Organization (NAFO), a regional fisheries management organization, has responsibility for management and conservation efforts of groundfish and shrimp in these waters.

By 1995, all major cod and flounder fisheries on the Grand Banks were closed, and many other fish species such as Greenland halibut and redfish had their catch levels sharply restricted. The recovery of almost all these stocks in NAFO Divisions 3L, 3N, and 3O (Grand Banks) has been very slow or non-existent. A notable exception has been yellowtail flounder, which was not depleted to the same extent as other stocks and which has rebounded to historic levels. At the same time, an increased abundance of some shellfish species such as crab and shrimp has led to the recent development or increase of these fisheries in Canadian waters on the Grand Banks. There has also been the development of an international shrimp fishery outside 200 miles on the Nose.

A similar pattern has been observed on the Flemish Cap, another shelf of relatively shallow water centred 120 nautical miles east of Canada's 200-mile limit completely within international waters (NAFO Division 3M). The minimum water depth is deeper than the Grand Banks at about 140 metres, and the average water temperature is generally higher than on the northern Grand Banks. The 58,000-square-kilometre area may have served as an important refuge for marine species during the last ice age.

Major changes in the ecosystem of the Flemish Cap took place during the 1990s and have continued until the present. The abundance of cod and American plaice has declined, as has their distribution range, while Greenland halibut have spread into the shallower depths of the Cap and there has been an increase in the abundance of shrimp. There is currently no directed fishery for American plaice or cod in Division 3M, with little improvement expected in these stocks in the foreseeable future. The decline in redfish stocks seems to have halted, but it is unclear whether the stock will be able to rebound. There are significant international fisheries for Greenland halibut and shrimp in the Flemish Cap area.

Even though there is a moratorium on fishing many species on the Nose and Tail of the Grand Banks and the Flemish Cap, these stocks can be taken legally as bycatch in the directed fishing of other species. The allowable bycatch, however, is only 5 per cent, and there are significant concerns that these limits are routinely exceeded. NAFO has expressed grave concern over the increase in catches of cod in Division 3NO and American plaice in Division 3LNO, currently at levels that will not allow these stocks to recover. The Government of Canada increased its monitoring activity in the NAFO Regulatory Area and its diplomacy efforts internationally in 2004 and has seen better compliance as a result.

For nearly a generation now, all Grand Banks fishermen have lived beneath the umbrella, or the shadow, of the 1977 act. Initially seen by the fishermen as a declaration of independence from foreign factory trawler fleets and a surety for protecting the vital Grand Banks cod fishery, its legacy has been neither freedom nor protection. It has created not only endless regulatory shoals and Sargasso Seas of paperwork, as fish landings dropped drastically since the early 1990s, but also ever-smaller quotas and more restricted fishing areas and permits.

How could this happen? Another example of an unintended consequence? In the 1960s, Canada was one of the first countries to declare a twelve-mile fishing zone and, in 1977, extended it to 200 miles. Meanwhile, many Grand Banks fishermen became incensed that the foreign fleet was killing the cod stocks. They turned to Fisheries Minister Brian Tobin, a man who instinctively knew how to play this card: blaming the foreign devils for over-fishing the Grand Banks would not only deflect blame from the Canadians, it would cement his image as a tough-guy fisheries minister, who would one day make a run at prime minister.

In May 1994, Canada gave itself the legal authority to seize foreign fishing boats outside its 200-mile limit in the North Atlantic. The law, enacted by both houses of Parliament, focused on an area about half the size of New York State, known as the *Nose and the Tail of the Grand Banks.*

"We're going to very clearly tell the vessels that we sight or target that they either move or be moved," Tobin said at a news conference in St. John's, Newfoundland, at that time. He said he expected Canada to start enforcing the law within thirty days, the time it takes to get the final regulations written and approved.

While the law was politically popular here, it drew fire, and continues to draw fire, from domestic critics and foreign governments. Some say that if Canada is able to extend its maritime jurisdiction to international waters, other nations might do the same and unravel years of negotiations on the Law of the Sea. At the time, Canadian officials, however, said both the new law and the government's stated rejection of international challenges to it were temporary measures in response to an emergency situation.

"Canada is acting purely because it is driven to act," said Tobin, chief architect of the law. "We have conducted diplomacy down to the last few pounds of fish. We either continue to talk about it, or we try to save those last few pounds and use them as a basis for restoring the stock."

Under the new law, any ship found taking endangered species was to be towed into port, its captain arrested and put on trial. If convicted, he could be fined as much as $360,000 and lose his catch.

The crackdown was aimed at "pirate" vessels, which the law defined as those either without international registry or flying "flags of convenience" and fishing just beyond Canada's 200-mile limit.

To avoid international conservation agreements, some foreign fishermen took to flagging their vessels in countries like Panama, Honduras, and Belize. These flags of convenience require little or no compliance with international rules.

3

Botched Science and a Rebel Named Ransom

"This is a crime beyond imagination."
—Ransom Myers

How's this for an insider's game played out in the Ivory Tower? The feminist revolution notwithstanding, fisheries science remains a mostly male occupation where grizzled PhDs slump over laptops inputting data, formulating algorithms, and analyzing results for computer models. Once upon a time, Canada's fisheries scientists basked in their excessively praised, international reputation as the best in the world; until, that is, the catastrophic collapse of the Grand Banks cod. Made in Canada, it is considered the world's worst ecological collapse, parallel only to clear-cutting in the Amazon rain forest.

At a press conference in St. John's on July 2, 1992, John Crosbie, Canada's flamboyant minister of fisheries and oceans (DFO), announced his government's decision to impose a moratorium on the fishing of the

Northern cod stocks. The clunky institutional prose of the news release sounded the death knell for a cod fishery that had defined the island and its people for 500 years—marking the death of rural Newfoundland and a subsequent nightmare for the minister, his government, and the fishermen.

Details of the moratorium were further explained the next day on a dock at Bay Bulls, just south of St. John's, where a clutch of angry fishermen, fuelled by union rhetoric and pints of lager, pushed and shoved the minister and his entourage. Seething with rage, the fishermen reverted to a long-held island tradition—they blamed Ottawa for the collapse of the stocks.

They had every reason to be pissed off. The closure, the single largest mass layoff in Canadian history, gutted the heart of Newfoundland. The moratorium was only supposed to last for two years, but has now been in effect for sixteen, leading to double-digit unemployment and the loss of an astounding 70,000 people (12 per cent of its population) to out-migration in the past decade. Nearly all the fish plants were closed, taking thousands of jobs, many residents, and a sense of identity from this island of 510,000 people.

However, had anyone chosen to listen, there had been a series of dire warnings.

The warnings had come from Newfoundland's inshore fishermen— the traditional small-boat fishermen who operate during the months when cod migrate inshore to feast on capelin. In 1982, the size of both the inshore catch and the individual fish making up the catch began to decline and fishermen accused the offshore fleet of fishing the stocks too heavily. Inshore catches continued to fall between 1982 and 1986; landings dropped from 113,000 tonnes to 72,000 tonnes during that time. Inshore fishermen became increasingly vocal about the perils of overfishing. The government dismissed their concerns. Federal fisheries information officer Bernard Brown described the government's attitude: "Essentially, the government was telling the inshore fishermen, who were creating all the uproar about the destruction of the stocks, that they didn't know what they were talking about."

The government's reluctance to listen to the inshore fishermen reflected the difficulty that a centralized bureaucracy faces when dealing with decentralized information. The Department of Fisheries and Oceans obtained most of the data that it relied on to assess stocks and set catch limits from the offshore fishery. It found it far easier to get information from fifty offshore trawlers owned by a few companies than from tens of thousands of widely dispersed small-boat fishermen using different gear in different ways. Information from the offshore fishery was systematic, uniform, and quantifiable, while the information from the small-boat fishermen was often anecdotal and could not be readily quantified or computerized. In the words of Information Officer Brown, "You just don't want to deal with that kind of messy information." Edward Sandeman, former director of the DFO's science branch, defended that attitude:

> There is a fundamental reason why, to a large extent, we ignored the inshore cod fishery. The reason being that it was extremely difficult to study . . . It was just too big an area to cover with the people we had . . . [T]he comments of the vast majority [of inshore fishermen] are self-serving and extremely restricted in geographical range. For the most part the majority of them have a litany of mumbo-jumbo, which they bring forth each time you talk to them.

In 1986, the Newfoundland Inshore Fisheries Association became more scientific. It commissioned three biologists to review the government's stock assessments. Their review criticized the government's sources of data, its statistical procedures, and its conclusions about the status of the Northern cod stock. It charged that the government, systematically interpreting uncertain information in the most optimistic light, had overestimated the fish biomass (the total weight of the stock) by as much as 55 per cent each year; as a result, the government had permitted catch levels that had prevented the stock from recovering from overfishing by foreign boats in the 1970s. True to its habit of rebutting rather than communicating, DFO dismissed the review as superficial.

Although the inshore fishermen sounded the loudest warnings, they were by no means alone, wrote Elizabeth Brubaker for a Hoover Institution Press study in 2000: "Scientists within the government also expressed uncertainty about the stock assessments upon which their political masters based catch limits. Their cautions about the unreliability of their data and conclusions, however, were often stifled by a bureaucracy intent on simplifying its findings for political or public consumption. The institutional structure of DFO could not accommodate ambiguity or uncertainty. Decision-makers wanted simple, unproblematic information."[1] As a result, scientists were frequently, in the words of Brian Morrissey, former assistant deputy minister of science, "drawing a firm line with a very unsteady hand."

Insisting on an appearance of unanimity, DFO always presented its stock assessments and its recommendations regarding allowable catches as consensus documents. Former employees have complained that the department's determination to have a single official opinion compromised the quality and effectiveness of its work. Stock assessment documents failed to reflect the full range of scientific opinion regarding the health of fish stocks. Senior bureaucrats and politicians accordingly set catch limits without access to information about the full implications of their decisions.

The demand for consensus did not just impede the decision-making process—it also limited intellectual discourse within the department. Former DFO biologist Jeffrey Hutchings described the department's tendency to reject thinking that challenged established positions: "It seemed to behave almost as a tribe, as tribal groups. It was group thinking and group action . . . and if you got someone from outside of that group analyzing what you've done, I think there was a tendency to downplay or, possibly, discount it."[2] The result could be disastrous. As Ransom Myers noted, "bureaucratic and authoritarian control, over scientific results, results in pseudoscience, not science. Such a system will inevitably fail and lead to scientific blunders."

In addition to minimizing uncertainty and creating the illusion of consensus, DFO scientists frequently put an inappropriately positive spin on ambiguous information. In assessing cod stock sizes, they tended to interpret ambiguity too optimistically, exploiting the "interpretive flexibility" of the data. Sometimes they did so to meet their own ends. Leslie Harris, chair of the 1989 Northern Cod Review Panel, commented on scientists' inclination to use data to confirm their projections: "I think our scientists saw their data from a particular perspective that the stock was growing at the rate they had projected, and the data were sort of made to fit the equation."

Other times, scientists selectively presented or interpreted data to meet the needs of their political masters. They understood that they derived their funding and authority from politicians who relied on their help to achieve political objectives.[3]

This understanding tainted the stock assessment process. One critic charged that "the actual dynamics of the CAFSAC [Canadian Atlantic Fisheries Scientific Advisory Committee] process shows it to be more a forum for projecting the political interests of the state into the scientific construction of reality than the other way around."

Jake Rice, former head of DFO's Groundfish Division, admitted that there were times when political realities prevented him and his colleagues from disclosing the full scientific truth: "Or you can only tell half the answer because the other half is still being debated in Ottawa for its political sensitivities. I, and no other scientist in the department that I know of, have ever been asked to lie. But we certainly have, at various times, been discouraged from revealing the whole truth. Every government has to do that to its civil servants."[4]

The tendency to ignore uncertainty and to interpret ambiguous data optimistically affected the political bureaucracy even more severely than the scientific bureaucracy. According to Brubaker, one DFO employee explained that although decision-makers did not falsify documents, "they optimized what they had": "The politicians and the senior bureaucrats would run away, pick the very best numbers, and come out and present them in the very best light. They would hide any negative notions, numbers, information, anything at all that took the gloss off what they had presented. Any attempt by anyone on the inside to present a different view was absolutely squashed."

John Crosbie admitted to sharing this tendency towards optimism: "We have opted for the upper end of the scientific advice always striving to get the last pound of fish."

The habit was a long-established one. The 1982 Task Force on Atlantic Fisheries, for example, relied on explicitly tentative scientific documents rife with warnings about their untestable assumptions. The warnings never made it into the task force's report, which overflowed with wildly optimistic forecasts of stock growth. The Northern cod stock, the task force predicted, would grow explosively; within five years, the allowable catch would increase by 75 per cent. These predictions encouraged further expansion of the industry, justifying unsupportable investments in harvesting and processing. "The government of the day," commented Jake Rice, "did no one a service to drop the qualifiers from the original scientific documents."

Scientists who were aware of and concerned about the government's selective use or overly optimistic interpretation of data could not express their concerns. Jeffrey Hutchings explained that public employees must not challenge the government's position: "For a government scientist to publicly disagree or publicly identify scientific risks or scientific deficiencies in the minister's decision is to publicly disagree with, and potentially embarrass the minister, and it's simply not allowed."

Suppression is inherent in the rules of the civil service. The federal government's communications guidelines advise ministers to "ensure that communications with the public are managed . . . in accordance with the priorities of government. It is not appropriate [for public servants] to discuss advice or recommendations tendered to Ministers, or to speculate about policy deliberations or future policy decisions."

The federal media relations policy authorizes designated spokespersons to speak to the media only on matters of fact or approved government policy.[5] Furthermore, the collective agreement covering DFO scientists specifies that "the employer may suggest revisions to a publication and may withhold approval to publish." Although the government defends this restriction on the grounds that it applies to policy issues rather than data, it cannot be unaware that rules limiting a scientist's discussions with respect to data make impossible any meaningful communication, since even the simplest facts often have policy implications.

The Department of Fisheries and Oceans' departmental guidelines go even further in its efforts to stifle employees' voices: inserted into the same category as fraud and assault, "public criticism of the employer" may be grounds for dismissal.

Such rules intimidated scientists concerned about the cod stocks and suppressed their valid, if controversial, findings. When the House of Commons Standing Committee on Fisheries and Oceans held an inquiry into the role of science in fisheries management, successive witnesses testified about flagrant intimidation. The witnesses were not merely disgruntled former employees. Steven Hindle, then president of the Professional Institute of the Public Service of Canada, the union representing DFO scientists, appeared on behalf of scientists throughout the department who feared the repercussions of speaking out themselves. He described a "climate of intimidation and mistrust" within the DFO, saying that the department suppressed data, ignored, or diluted scientific advice, prevented scientists from publishing or speaking publicly about their findings, and threatened the career advancement and even the jobs of dissenters.[6]

As the cod disaster brewed, DFO did not merely control the information flowing from the department to the public, it also limited the flow of information within the bureaucracy, preventing critical knowledge from reaching upper-level decision-makers. Several instances involved the Canadian Atlantic Fisheries Scientific Advisory Committee (CAFSAC), which provided information on stocks to the Atlantic Groundfish Advisory Committee (AGAC), the body that advised senior management about catches. In 1990, CAFSAC's discussion of a conservation-based catch limit for 1991 did not even make it onto the AGAC's agenda. Thus, those who ultimately chose a limit of 190,000 tonnes may not have known that the option of 100,000 tonnes would have allowed for a sustainable catch.

A similar omission occurred on July 2, 1992, the first day of the Northern cod moratorium when the chair of CAFSAC's groundfish subcommittee was scheduled to make a presentation to AGAC. He intended to give AGAC an overview of all of Canada's Atlantic cod stocks that would indicate that, because stocks other than the Northern cod were also being overfished, their levels, too, were declining. Apparently senior officials did not want AGAC to hear that the other cod stocks could be in serious trouble, so they cancelled the presentation. Jeffrey Hutchings later speculated that the cancellation may have been motivated by the fact that findings in the presentation were inconsistent with the deputy minister's announcement, two days earlier, that the troubles in the northern cod fishery were unique.[7] Of course, it soon became obvious that stock declines were the rule rather than the exception. By the following fall, DFO had closed the cod fisheries off the south coast of Newfoundland and in the Gulf of St. Lawrence.

"This is a crime beyond imagination. During this year delay, 70 per cent of the remaining cod were removed and this caused a much greater collapse in the rest of eastern Canada than was needed. We could have stopped fishing then. It was a direct decision, a bureaucratic decision, to suppress the information," said Ransom Myers, a population biologist in the Canadian DFO in St. John's, at the time.

Scientists became an easy target for tough-minded Newfoundland politicians. Deflecting blame from the incompetence of his own ministry, Crosbie pointed his finger squarely at the "egg-heads" in the laboratory. They had botched the science that led to the moratorium.

Today, the mystery of the Grand Banks cod collapse has crossed over into the realm of public discourse. Many have questioned whether fisheries science—in particular its forecast models—can rightly be considered science at all. Science, it seems, is not as reliable and as precise as we'd like to believe. There is, after all, the problem of the whys and wherefores of scientific knowledge. Scientific knowledge, it turns out, is far more than bringing proof into the laboratory. On the Grand Banks and in federal offices along Water Street in St. John's, science is also a social construct in which many outside pressures and social issues have to be dealt with before knowledge is ultimately created.

How this happened at the DFO is detailed exhaustively by Alan Christopher Finlayson in his 1994 book *Fishing for Truth*. Government pressure on scientists for high total allowable catch (TACs) was heavy because the main industry in the province was built on the slippery back of one resource: cod. Despite uncertainty in the process, the scientists felt compelled to come through with larger estimates than they were sure were safe. Even when doubt was raised later on, it was swept aside due to socio-economic reasons. The resulting overestimates led to the TACs being set too high, and government market interventions and the capitalist ethos drove fishermen to go out and fill those TACs.

According to geologist Orrin Pilkey and environmental scientist Linda Pilkey-Jarvis, botched science was, indeed, at the heart of the Grand Banks cod collapse. The two are the authors of *Useless Arithmetic: Why Environmental Scientists Can't Predict the Future,*[8] a book that explains why the mathematical models used by scientists and governments are fundamentally flawed and can lead to unwise, if not disastrous, environmental policies.

Useless Arithmetic devotes an entire chapter to the demise of the Grand Banks cod fishery, pointing out that the key difficulty in the fisheries scientists' work lay in the assumptions they made in trying to estimate the cod population. Since it's virtually impossible to consider the whole ecosystem and all the complex interactions of various species, climate, and other factors, fisheries scientists generally focus their models on a single species, which leads to inherently unreliable results.

In assessing the stock size of the Grand Banks cod, scientists based their calculations on the age distribution of individual fish in the fish population. The mathematical model tells the number of fish that survived to a catchable size and from there scientists determined the size of the cod stock, which was then used to set the TAC. To determine this amount, scientists had to assume a "reasonable" mortality rate for the fish, which, in reality, can vary greatly.

Pilkey and Pilkey-Jarvis' book also calls into question government field sampling methods, recalling the case of the U.S. National Marine Fishery Service, which confessed in 2002 that its New England fish population estimates over the previous two years had all been wrong due to their use of uneven lengths of trawling cable. This error in population estimates led to the unpredictable behaviour of the trawling nets used to catch the fish.

Another scientific assumption that Pilkey and Pilkey-Jarvis challenge is that the size of the adult fish population is directly related to how many young fish survive into adulthood every year: the more adults, the more fry. This may sound logical, say the authors, but the factors that impact larvae survival at their time of hatching are very different from the factors that affect the fish at later stages of development. Therefore, more eggs don't necessarily equate to more adults.

Pilkey and Pilkey-Jarvis also cite other scientists' findings that further complicate the forecast models. For example, they refer to Daniel Pauly and Jay Maclean's 2003 book, *In a Perfect Ocean*, which reminds readers that the fishing industry's record is far from spotless. Juvenile or smaller fish may be wastefully discarded by fishers; some fish are caught but remain unreported, while others are simply caught illegally. All this activity can have a significant impact on the fish population.

One more factor complicating the fisheries science model is offered by Oregon marine ecologist Mark Hinson and his colleagues, who determined that fish size also matters when estimating a population. Larger fish not only produce more eggs, but the larvae hatched from their eggs are hardier than those from smaller fish, and thus have a better survival rate.

"Single-species models can't work and protect the entire ecosystem, but single-species models are really all we use," conclude Pilkey and Pilkey-Jarvis in their book. "It seems as though the more we know about fisheries, the less we know. Each step in the direction of understanding ecosystems reveals more and more complexities, and in any complex system in nature we can never obtain quantitative modeling answers at the level of accuracy that society needs. Society seeks an answer through fish mathematical models, but it can never get that answer from them."

Since publication of the Pilkey book, other fisheries scientists have weighed in on the Pilkey thesis. Daniel Pauly says Pilkey's cod analysis is sound, but warns that all modelling is not flawed, that it is a necessary tool, used to predict, for example, global warming. His colleague at the University of British Columbia, Carl Walters, who in 1997 co-wrote a technical paper on the same subject also holds up a caution: "The scientific screw-up in the cod case was only partly from using a bad model for 'nature' (they ignored a very simple ecological thing, namely that recruitment was falling as the spawning stock declined), but more importantly they misinterpreted field data on trends in catch per effort. So, if Pilkey argues that good field data is the answer to such problems, he could be way off the mark," Professor Walters wrote in an email.[9]

The problem in the Grand Banks was further complicated by deeply held beliefs about the abundance of cod. After all, the unprecedented richness of the cod stocks fuelled repeated military conflicts between nations for the control of access to these fishing grounds. Such abundance was also the principal motive for the European colonization of the otherwise barren and inhospitable land now known as the province of Newfoundland and Labrador. It's not entirely surprising that, even in the face of contrary evidence, the attitude throughout the fishery, including scientists, was that the cod was an abundant, never-ending resource.

Wrong, as it turned out, and fool-headed. By 1992, the Grand Banks cod had been driven to the brink of commercial extinction. The rise of modern ice-reinforced trawlers in the 1960s and 1970s hadn't helped. When, in 1977, Canada extended its authority to 200 miles, there followed a "gold rush" sanctioned and paid for by Canadian politicians. A new day for Newfoundland and the beginning of a fifteen-year assault on the cod stocks, which scientists and managers proclaimed, year after year, were doing just fine. Under the surface there festered unresolved structural problems with fisheries science—and cod stocks in serious decline.

For more than a century, the relationship between science and federally funded fisheries had been a struggle. Scientists wanted to stay at arm's length from government and its mandarins who wanted control. The final resolution came in 1973 when the Canadian government, anticipating the linking of foreign policy considerations to its greatly expanded management responsibilities of a 200-mile limit, simply eliminated the last vestiges of the Fisheries Resource Board's (FRB) independence by an act of Parliament.

Science had a new role. It was expected to help mitigate the "boom and bust" cycles that had plagued the Grand Banks cod fishery throughout the past. But, after Canada declared its 200-mile limit, governments in Ottawa and Newfoundland stumbled over each other in a rush to deep water, sending out icebreaking dragger fleets to haul in the abundance of Northern cod where the foreign draggers had once worked.

A brief period of euphoria descended on the province. This was to be Newfoundland's chance at last. Its inhabitants were poor, unemployed, and still living, some of them, in outports that could not be reached by road. Foreigners had shown just how many fish could be caught; now those fish would belong to Newfoundland. The government deliberately encouraged the expansion of the offshore fishery.

With an investment of hundreds of millions of dollars of taxpayers' money, Newfoundland reconfigured National Sea Products of Halifax, Nova Scotia, and Fishery Products International of St. John's, Newfoundland. The boards of directors were filled with political appointees; former civil servants took executive positions; almost 50 percent of the total available fish quotas in Eastern Canada were given to these new super-companies. This privatization of a previously common-property resource essentially created a swimming-fish inventory for the dragger corporations that they could harvest at their leisure.

Hundreds of millions of government dollars were pumped into building and refurbishing a massive corporate dragger fleet. It took a few years to establish, but by the winter of 1979, and every winter since, Canadian draggers have gone onto the offshore Northern cod spawning grounds and towed huge net bags through the dense masses of spawning fish.

Dragging is like towing two D9 tractor blades, linked with 150 feet of very heavy chain and rollers, back and forth across every inch of the fishing grounds. If an oil company had proposed such a process without an environmental assessment, they would, quite correctly, have been threatened with a good boiling in oil. But "fishing," conducted by "quaint fisherfolk," was somehow outside the realm of environmental concerns and regulation.

Yet, these "fisherfolk" hauled gigantic bags of fish, hundreds of feet to the surface, by huge winches in a matter of minutes. The shock of such an experience killed every living thing. Once onboard, the fish were ruthlessly culled. Anything that didn't maximize corporate profits was flushed overboard. It will never be known how many hundreds of millions of fish were destroyed this way. A 1986 government report estimates that 16 million cod were flushed overboard by National Sea and Fishery Products International draggers in one eight-week period.

The slaughter went on year after year and only stopped in 1992 because Fishery Products International and National Sea found that when they went to the spawning grounds there were not enough fish to make a profit—hence the moratorium.

In 1977, in the midst of the glee arising out of the vast potential of the newfound offshore fishery, DFO biologists were asked to forecast how fast the cod stock would grow now that foreign factory trawlers weren't gobbling it up. Amid the euphoria, the stock assessors must have felt considerable pressure.

Some senior officials did not expect, did not appreciate, did not want to hear the dire warnings that the stocks were collapsing. To predict the potential growth of the cod stock, researchers had to estimate how many young fish would be recruited to the fishery each year. They made an important assumption: that future recruitment, after 1977, would be the same as the average during the 1960s and 1970s. During all that time, though, the stock had been continuously declining as factory trawlers mowed the banks. In fact, by 1977, Myers and his colleague Jeffrey Hutchings' analysis of the data show that the number of spawning cod off Newfoundland was down by 94 per cent from what it had been in 1962. This was a message repeated by the small-boat fleet of the inshore fishery as well. Warnings from both groups were persistently avoided.

The voices continued. Catches were down and cod were smaller. Few scientists were listening; but Ransom Myers was an exception.

Considered one of the world's pre-eminent scientists, Myers, who died in June 2007, of brain cancer, was an iconoclast who put the sustainability of the cod stocks ahead of his career. He worked at DFO from 1984 to 1996, when he accepted an academic post at Dalhousie University in Halifax.

Although he became known for a series of papers in science and nature magazines, it is probably his work on the recruitment problem that remains his greatest legacy. He used the mathematical skills he acquired through earlier degrees—a BSc in Physics from Rice and an MSc in Mathematics at Dalhousie where, in 1994, he also received his PhD. He performed a meta-analysis to normalize data so it was comparable among different species. Until his work was published, it was commonly held that overexploited fish stocks, such as the Grand Banks cod, could easily rebound from depletion. In 1997, Myers put to rest this great myth of fisheries science, which so often led to overly optimistic scientific proclamations.

While he was conducting his research, the Grand Banks cod collapsed. Myers published evidence that excess fishing pressure was the cause of the collapse—not seals, cold water, hot water, or other excuses from an agency who, by caving into industry pressure, had failed to protect King Cod and the province whose economy and livelihood depended on it.

According to Myers, the science, as well as the entire fisheries system, was deeply flawed. In return for his truth-telling, DFO issued a formal reprimand, but it was soon after this that he quit the department and accepted a position as the chair of Ocean Studies at Dalhousie University.

In 1997, at a Standing Committee on Fisheries in Ottawa, Myers explained just how bad the science was during a review of the role of science in fisheries management. In his patented plain-speaking style, Myers told the committee that fisheries science had been trumped—corrupted by upper-level bureaucrats, answering to politicians in Ottawa and St. John's:

> Such a system results in pseudoscience, not science. Such a system will inevitably fail and lead to scientific blunders. There was suppression of research within DFO. Data were secret and not allowed to be analyzed.
>
> Tagging data that would have shown there was very high fishing mortality were kept secret by the department. Research surveys that could have been looked at in more detail were kept secret.

Myers cited the example of an analysis carried out by George Winters in 1986 showing that fishing mortality was much higher on Northern cod than had been estimated. "This paper was directly suppressed by the director of the lab at the time," he said, "because the information was unwanted." There was intimidation. "Researchers who came up with results that were not bureaucratically acceptable," Myers said to the committee, "were intimidated by a variety of means. In such a condition, when one is being intimidated by their bureaucracy, it is impossible to carry out open-minded research."

It was against this backdrop of intimidation and flawed science that Newfoundland's fishermen prospered for a short time while the cod continued to die at a dangerous rate. By chance, the years from 1978 through 1981 really did turn out to be a time of relatively good recruitment—although not as good as the DFO had predicted. During those years, with the foreign trawlers gone, the cod stocks off Newfoundland actually increased. But so did the ability of Newfoundlanders to catch them.

Boosted by federal subsidies and by DFO's optimistic unprecedented clip, Newfoundlanders had otter trawlers now—ships large enough to permit offshore fishing, even in bad weather. Those ships had modern navigational and sonar equipment that allowed fishermen to locate the large spawning schools and to sweep them up. With the goal of a 350,000-tonne harvest, DFO increased its catch quota several times during the early 1980s. But the cod were simply not there in the numbers DFO was claiming. DFO had become a cheerleader, not a manager.

By the mid-1980s the problem was apparent to outside critics and was even reflected in the DFO's own data. Fishermen, as it turned out, had been harvesting a much higher percentage of the cod stock than DFO had thought in the 1970s and 1980s—so high that the stock could not be growing at the rate DFO had projected.

The inshore fishermen, using centuries-old skills of local knowledge, noticed few cod in their fishing holes, and those that they did catch were smaller. They became so concerned they took DFO to court—but lost.

By 1985, Myers and his colleague Hutchings warned DFO yet again, only to be rebuffed once more. The Grand Banks cod stocks had begun their long slide into catastrophe; a slide that couldn't be stopped. By 1989, the political and economic momentum behind an expanded Newfoundland fishery was too great. Not wanting to throw thousands of people out of work, the minister of Fisheries and Oceans rejected his stock assessors' advice to cut the 1989 cod quota all at once by more than half, to 125,000 tonnes. Instead, he cut it by a tenth.

"The analysis of them was completely botched," said Myers at the time. "So you were already taking out too many fish, but because of the error you were taking out tremendously too many.

"And toward the end, as the cod population declined, people tried to maintain their catch rates to maintain their incomes. So, they fished harder. Inshore fishermen were going in small boats 100 miles offshore and setting bottom gill nets—they were going way the hell out under incredibly dangerous conditions. That caused the fish population to go down more quickly, which caused the fishermen to fish harder," Myers told the committee years later.

Meanwhile, it was clear to Myers and his team that there were not many older fish of spawning age; of course, there weren't many younger fish, either. There were lots of people who thought fishing mortality was too high and should be reduced. But no one suspected the magnitude of what was happening. Instead of being two and a half times what was desirable, the mortality was five times too high. When the fishing mortality is that high, the stock can expect to collapse very quickly. In 1991, by Myers' and Hutchings' reckoning, Newfoundland fishermen caught more than half the cod living in their waters, some 180,000 tonnes. In December of that year, DFO recommended that they catch the same amount in 1992.

It never came to that: in July 1992 the minister was forced to close the Grand Banks fishery when the stocks collapsed. By then there were next to no cod of spawning age, seven years or older, left. There was just 22,000 tonnes' worth—less than a quarter of what there had been in 1977, after the factory trawlers had done their worst. Chaos ensued. Consternation and angry voices were raised throughout the province. Then came the talk of conspiracy and betrayal accompanied by yet another round of Ottawa bashing.

In 1982, and again in 1987 and 1989, generally perceived crises in the fishery brought about the formation of federally sponsored task forces to investigate.

The first was known formally as the "Task Force on the Atlantic Fisheries," and informally as the "Kirby Commission," referring to its chair, Michael J. L. Kirby. The group formed in 1987 was named the "Task Group on the Newfoundland Inshore Fisheries" (TGNIF) or the "Elverson Commission," chaired by Dr. D. Lee Elverson. The 1989 group was known formally as the "Independent Review of the State of the Northern Cod Stock," and informally as the "Harris Commission"; its chair, Dr. Lisle Harris, was then the president of Memorial University of Newfoundland.

The three reports were different in almost every respect, sharing only a general sense of crisis in the fisheries and a federal mandate. At the heart of the matter was the deceptively simple question: "How many fish are in the sea?" Each report was respectful and diplomatic to a fault. Each, in its own way pointed out the high degree of uncertainty in the scientific assessment of fish stocks, particularly migratory species such as Northern cod.

All the reports gave brief descriptions of the historic bounty of the Grand Banks cod and, how for 400 years the Banks had provided food for fishing fleets from around the world. Each report referenced how the industry dramatically changed from the small boat dories to high-tech industrial trawlers in the 1950s and 1960s, dividing Newfoundland fishers into two distinct groups: the traditional inshore fishers and the offshore fleet.

There are only two things we know with certainty about the Newfoundland fishery. The first is that prior to the introduction of draggers, a multi-national fishery with no quotas, rules or regulations, using anchored nets and lines of baited hooks, caught more and larger fish than we have in the past twenty years, without depleting the stocks. Each year vessels from Europe, Canada, the United States, and Newfoundland caught as much fish as they could. For 450 years there was a large standing stock of fish out of which a huge surplus was harvested.

The second thing we know is that the introduction of the dragger in the 1950s destroyed the massive stocks of fish, which had supported one of the largest fisheries on the planet for 500 years.

A curious publication from DFO was released in 1988: *The Science of Cod.* A patronizing attempt at spin, it touted the excellence of the very scientists who had failed on the Grand Banks: "Scientists are methodical. They value only what they can measure. Guesses, hunches, impressions, rumours, pet theories, likes and dislikes—all these things the rest of us find so absorbing must be avoided by a scientist. To be any good as a scientist or advisor, the DFO biologist must be neutral, objective, and professional."

The facile publication is filled with platitudes that render it a collector's item. Only in Canada would a government attempt such tomfoolery:

> Scientific knowledge is like a huge pool, which belongs to everybody and which grows as new knowledge is added. But not just any new information is dumped in. Scientists are cautious, skeptical folk, and each new contribution to the pool of knowledge is closely examined by other experts in the same field.
>
> In this process, sloppy work soon gets discarded. And the same strict standards apply whether the information is some new discovery or just raw data.
>
> In the case of stock assessment, the peer review process is complex. Each of the steps involved is a safeguard against poor research or hasty conclusions.

This publication was utterly false; as events were proving out on the Grand Banks.

Over time, DFO scientists published a type of defence in a series of academic papers that were written in a style slightly less than PhD English. In the papers, the scientists hinted at mistakes without ever taking responsibility and provided mountains of obscure research that produced a mouse of a thesis.

What is true is that, despite algorithms, computer modelling, and data collection, the present understanding of the fisheries management process, founded upon biological science, is wrong. Fisheries managers do not now and probably never will know enough about fish and their ecosystem to construct enough facts to support agreement and cooperation. Garbage in, garbage out.

Meanwhile, news of the Grand Banks cod collapse travelled across the country and around the world. Out on the Pacific Coast (see chapter 9), managers adopted eco-system risk-averse regimes and opened up the process to NGOs, environmental groups and, increasingly, the general public.

In today's Newfoundland, nothing has changed. While it's estimated that Grand Banks cod stocks are 1 per cent of what they were in 1977, and despite new warnings about the shrimp and lobster fisheries, many Newfoundlanders, long accustomed to a hardscrabble life and the "catch-and-kill" sense of survival, react to these latest warnings with a yawning complacency. Bring on the next, they say, and reverting to form, blame Ottawa.

Federal scientists did not catch the fish, but it was botched science that led directly to overly optimistic catch levels from 1977 to 1992, setting the table for the slaughter to come. Little wonder that scientists and many DFO staff are held in such deep derision to this day. Fishermen have long memories; that infamous day is remembered in books, and poems, and drinking songs. Botched science was the reason the offshore fleet idled, and 40,000 people were put out of work, at a cost of billions of dollars to Canadian taxpayers. And, King Cod has never come back.

4

Foreign Devils

"For a major fishing nation, we have the worst fisheries management in the world."
—Gus Etchegary

In 1992, Canada's federal government introduced Draconian curbs on fishing off Newfoundland, after admitting that stocks of cod in the North Atlantic were dangerously low. Many Newfoundlanders blamed, and blame to this day, the European fleets for this plunder. The Newfoundland fishing fleets knew that the Grand Banks were severely overfished during the 1970s. After Canada started to police fishing within its 200-mile offshore limit in 1977, the stock began somewhat to recover, but throughout the 1980s, fishermen could not find enough fish to fill the quotas.

Prior to 1989, scientists at the Department of Fisheries and Oceans estimated the state of the stocks by averaging the findings of the scientists' own research surveys and commercial estimates. DFO's figures came from samples taken from research ships, while the commercial figures were based on the amount of fish caught per unit of fishing effort. The commercial figures were consistently higher than the DFO's. This, the DFO admitted in 1989, was not because the stock of fish was increasing but because fishermen had acquired better technology and could catch more fish per unit of effort. Its own figures, which are based on random samples rather than a concerted effort to find fish, were in fact more accurate than the commercial ones.

So the DFO scientists recommended that the 1989 quota of 235,000 tonnes of cod should be cut to 125,000 tonnes. John Crosbie, minister of fisheries and oceans, called the scientists' advice "demented." He argued strongly against cutting quotas because of the "economic, social and cultural effects" such cuts would have in Newfoundland, where fishing is the main occupation.

In 1992, Crosbie finally, and reluctantly, accepted the scientists' advice after fishing companies in Newfoundland agreed that cuts were needed. Crosbie was quick to blame the shortage of cod on overfishing by Spanish and Portuguese boats. Catches outside Canada's 200-mile limit—which includes some of the Grand Bank—are set by the Northwest Atlantic Fisheries Organization (NAFO). In 1986, under pressure from under-employed fishermen in Spain and Portugal, the European community started setting its own, higher, catch limits.

There were clashes about other species, too. In 1995, a small war erupted over turbot, known in Europe as Greenland halibut, taken by Spanish vessels in the Grand Banks' famous "Nose" and "Tail," which are located in international waters. Canada seized one Spanish trawler diverting it to St. John's, Newfoundland, and cut the net from another. Spain sent patrol vessels into the area to protect its fishermen and the rest of the world sat back to await a shootout on the high seas.

What they got was a high-stakes public relations battle. Canada laboured to offer legal justification for its actions and sought to depict itself, through its telegenic fisheries minister, Brian Tobin, as a kind of high-minded conservation vigilante. They displayed nets from the commandeered trawler claiming they were too finely meshed to be legal and alleged evidence of a "double log" and a "secret hold" for illegal fish aboard the craft. Canada said that Spain was mindlessly flouting the law and endangering stocks by pulling in fish not yet mature enough to reproduce.

The European Union, of which Spain is a part, leaped to its defence. The Spanish ships had every right to be there, officials argued, adding that the so-called "evidence" did not hold up and there were no minimum-size restrictions on turbot. In fact, far from seizing the moral high ground, they said the Canadians had broken international law, which European fisheries commissioner, Emma Bonino, likened to an act of "piracy."

Behind the furious faxes, press conferences, threats, and a lawsuit filed by Spain at the International Court of Justice located at the Hague, lies an indisputable fact: fishing is big business and the price of fish goes up as stocks go down. New techniques, using satellite tracking and specially designed nets, are more voracious. Fleets keep getting bigger; 30 per cent of ships over 100 gross tonnes are now fishing vessels.

One man, Gus Etchegary, blames the foreign devils for the collapse of the Grand Banks cod and has never stopped crusading against them. Etchegary, who began his career in the fishery in 1945, is the former chairman of the Fisheries Council of Canada and former president of Newfoundland's largest fishing and processing company, Fishery Products International (FPI). His extensive experience in fisheries led to an appointment to the NAFO.

A household name in the province, Etchegary has been a regular on talk-show radio for years. His stentorian rants are as familiar to the locals as ice and snow. His message, fine-tuned over the past four decades, resonates in the kitchens of the island. In his church, he speaks to thousands of converts, many of whom live on the eastern and southern shores of Newfoundland and they tell of, prior to the 1977 extension, a long blinking line of lights as the huge factory trawlers set up shop on Grand Banks, night after night: an illumination that brings the blood to boil.

In the summer of 2007, Etchegary was interviewed by a visitor from the West Coast and with, little time for pleasantries, the leather-lunged Etchegary hit first and kept on punching as he talked incessantly.

"I don't know who you've been talking to," he said. "You'd better listen to me."

So, I did. Since his sixty-minute rant certainly allowed no interruption, taking notes was the only reaction of those listening as the crusader punched home his take-away message: the Grand Banks have been overfished to near extinction by the foreign fleet—the Russians, Spaniards, Icelanders, Portuguese, and others. The foreigners did it. Full stop. Period.

It was only later that the visitor would learn that this was a practised presentation, one that had been honed over fifty-five years. Five years earlier, Etchegary had delivered the long version in his presentation to the Standing Senate Committee on Fisheries and Oceans in Ottawa, one of the many investigations to delve into the mystery of the Grand Banks collapse. Etchegary was in fine form on December 3, 2002, as chairman Senator Gerald J. Comeau opened the proceedings.

The senators, rightly, wanted to learn—as do Canadians from coast to coast and people around the world—why did one of the world's most productive fishing grounds collapse? Why were there seemingly plenty of cod one year and none the next? What caused this shocking collapse?

As explained in the senators' briefing notes, the Newfoundland Grand Banks, off the east coast of Canada, were world famous as amazingly productive fishing grounds. The first European explorers described the waters as being so full of cod that a basket lowered into the water would emerge completely full of fish. In the centuries that followed, abundant fish stocks drew many people to Newfoundland. Small inshore boats took sustainable amounts of cod for centuries up until the 1950s.

Foreigners are to blame, Etchegary told the senators. The foreign fleet plundered the Grand Banks cod resulting in the cod collapse and subsequent devastation in Newfoundland. The outports were especially affected as they were "now facing horrendous economic and social problems in the communities, and so they felt that they should come together and focus on this issue."

He continued: "The East Coast of Canada has been very badly affected by overfishing and by the lack of management. For a major fishing nation, we have had the worst fisheries management in the world. There is absolutely no doubt about that. Comparisons of Canada's performance with that of Iceland, Norway and many other countries that we are aware of, show that we have much to be concerned about."

Etchegary's argument was two-fold. Not only did the foreign fleets overfish the East Coast, but the DFO did nothing to stop it. In 1957, he explained, "there were 600 large factory vessels and 44,000 European fishermen fishing off the East Coast of Newfoundland and Labrador. That went on for thirty-five years unchecked—no monitoring, nothing."

From his personal experience, Etchegary recalled a conversation with a veteran skipper who once worked on a British factory vessel. "He told me that on their first voyages to Labrador in 1953, they were up to their ankles in spawn from fish that were caught off Labrador in, believe this, three and four feet of ice . . . These vessels were fishing for very high concentrations of spawning fish. Therefore, the downside was the result of thirty-odd years of uncontrolled fishing."

Etchegary doesn't, however, believe the DFO was simply ignorant of the situation. "The scientists and fishermen who were living with this knew it was happening. They knew that there was going to be a collapse in five, ten, fifteen, twenty years."

Of course, he granted, Canada couldn't very well lead the country to war over fish, but still, the government needed to stop compromising and taking an "impish attitude" in international negotiations.

Newfoundland and Labrador had already lost 50,000 people since 1993, Etchegary noted, and would surely lose even more if another moratorium were announced in the months to come. "The fishermen will not stand by in Newfoundland and Labrador and accept another moratorium when, at the same time, there are thirty, forty, fifty, sixty, seventy vessels and sixteen or seventeen nations fishing on our doorstep, landing that fish in our ports and shipping it back to their home countries. Or, they are shipping it to China for further processing with ten-cent-an-hour labour, providing them with all the goods and services they need to facilitate their operations so that it can be done as cheaply as possible. That, combined with the subsidies they are getting from the European Community, in many cases, will cause a reaction in the spring that has never been seen before.

"Why are DFO and its officials not alerting the senior cabinet ministers and the Prime Minister to this situation?" Etchegary wondered. "It is beyond me, and beyond a lot of people who understand. That is the kind of thing we are facing."

As he recalled, senior scientists "with tremendous corporate knowledge" had recommended not closure, but reduction of effort in order to rebuild the stocks. "They were ignored," he said. "Not only were they ignored, but a federal Minister of Fisheries of the day put in place a financial subsidy instead."

The subsidy encouraged Canadian offshore trawlers to return to the Labrador fishery, and in 1978, 150 Canadian trawlers were in the waters, "attacking the young, concentrated cod for spawning." They caught vast quantities of small fish, 80 per cent of the catch so small that they yielded the lowest return in the market.

That went on for five years before it was stopped. Then the moratorium was declared in 1992, and the fishery "has not recovered since because of the foreign overfishing," said Etchegary.

"Foreign overfishing is a real problem," he told the senators. "There is no question about that whatsoever . . . That is important, and foreign overfishing must be stopped somehow. However, the biggest challenge that we have in Newfoundland and Labrador is not with the foreigners, it is with Ottawa. That is where our problem lies."

He dismissed any claims that DFO didn't realize the extent of the situation given all the information they had been provided. "We were urging that we could not live with a 200-mile limit. We had to have an extension of jurisdiction. I will never forgive some of the senior bureaucrats of the day with whom we were talking so that they could inform their bosses and the politicians," he said.

To put it bluntly, he said, "DFO does not understand and does not have the interests of the East Coast of Atlantic Canada in mind."

Where did it all start to go wrong? Etchegary recounted that the transition from salted fish to the fresh-frozen fish industry began during the war and grew quickly, with no one at the time worrying about conservation or resource management. Looking back at historical records of how many fish were caught since 1875, said Etchegary, there was a consistent take of about 200,000 tonnes. "That would have continued indefinitely at that level. However, when that figure increased to one million tonnes then, of course, the crash took place. New technology was introduced and it was indiscriminate, in that there was no monitoring. There was a huge foreign fleet that dwarfed the total Canadian fleet in comparison."

Perhaps the disaster would have been preventable if the DFO had actually gotten its hands dirty at the fishery, Etchegary added.

Over the years, I have not met one person in a senior bureaucratic position with the Department of Fisheries and Oceans who knew a thing about the fishery. I am sorry to say that, but not one knew about the fishery. The department's officials just did not have the experience and the background. In Iceland and in Norway, officials who put the regulatory and enforcement measures in place are in the same harbour—in the same fishery. They interface with the industry day after day, week after week, and month after month.

You cannot run a fishery from a building in Ottawa, 1,800 miles away. Certainly, people who do not understand the economics of the fishery cannot do it. Some of those people do not have a clue.

"The moratorium has been an absolutely useless exercise," he insisted. "Many people are coming to the conclusion that this is cultural genocide."

Etchegary would have worn the senators down. To a visitor in the summer of 2007, he was an over-revving engine of a man who just wouldn't quit. Like a veteran politician on a perpetual stump, he hit first and kept on punching his key message: foreign devils—and DFO's failure to control them—were the primary cause of the cod collapse.

But beyond the bombast, can this possibly be true? Does he protest too much? Especially when considering that he worked in the fish processing industry, where, as president of FPI, he dispatched a fleet of draggers out to that watery killing field. With weighted nets up to 100 metres in length, the draggers bulldozed and tore up the spawning and rearing grounds of the fish whose passing he purports to lament today.

Etchegary is wrong. New data, uncovered and put together in the summer of 2007, offer documented proof that the killing machine that plundered and pillaged the Grand Banks, driving the stocks to near extinction, was made right here in Newfoundland.

5

Draggers and Cod Death

"Federal unemployment insurance is the life blood of rural Newfoundland."
—John Crosbie

In 1977, Canada extended its jurisdiction over fisheries from 12 to 200 miles, capturing most of the continental shelf, but not the Nose and Tail of the Grand Banks or the Flemish Cap. This was the biggest party since Confederation, for a naturally celebratory people. Boozy cries ricocheted around the bars that line St. John's George Street, as well as the tiny pubs in outports of every corner of the province: "It's ours now, boy. Let's go cod fishing."

"It was a gold rush kind of mentality," said one bureaucrat at the time, while his colleague suggested the new 200-mile limit created a "bonanza attitude, El Dorado again."

Richard Haedrich, a biology professor at Memorial University of Newfoundland who worked as a biologist back then, said: "The idea was that the streets were paved with fish and now that the Europeans were gone, it would come to the Canadians."

Almost immediately, politicians in St. John's and in Ottawa, desperate for any new idea that could alleviate the grinding poverty and high unemployment in Canada's have-not province, joined the frenzy, falling over themselves to draft and rubber-stamp a series of bills to provide economic subsidies—grants, loans, tax exemptions—to build a new Canadian mid-size trawler fleet.

How did the Atlantic cod, which once seemed a limitless resource, suffer such devastation? Looking back across the centuries, cod were probably fished in Newfoundland waters as early as the late 1400s. Since the voyages by John Cabot (1497, 1498) on behalf of the English Crown, and by Gaspar Corte-Réal of Portugal (1500, 1501), present documents record the presence of fishermen from France and England in 1504, from Portugal as early as 1506, and the Spanish Basques from the 1520s.

France and Portugal established shore-based fisheries in Newfoundland in the early 1500s. The shore fishery constituted the harvest of fish from near-shore waters and subsequent drying of split cod either directly on pebble beaches or on shore-based wooden stages by a combination of sun, salt, and wind. The most heavily used location for shore fishing—by Portugal, France, England, and the Spanish Basques—was along the east coast between Cape Race and Cape Bonavista. The French also fished extensively westward from Cape Bonavista to Quirpon at the tip of the Northern Peninsula (this coast was known by the French as the Petit Nord), and then along the Labrador and Newfoundland shores of the Strait of Belle Isle. The French, Portuguese, and Basques also fished along the south coast, primarily from the southern tip of Burin Peninsula near Lamaline east to Cape Race.

According to some researchers, reconstructed catch and effort data suggest that there were certain signs of overfishing as early as the mid-1800s. The twentieth century brought the greatest push in the fishing industry. Rising prosperity and widespread refrigeration technology meant an increased demand for frozen fish, which fundamentally changed the groundfish industry. Large companies combining fishing and processing activities came to dominate the East Coast fisheries, building up fleets of trawlers that could catch almost as much groundfish as thousands of smaller boats.

All this pressure on the stocks eventually led to overfishing. Although the Atlantic fleet multiplied its capacity five-fold from 1959 to 1974, the actual cod catch fell as stocks were depleted.

In 1977, when Canada extended its fishing limits to 200 miles, pushing out foreign vessels, scientists projected strong growth for the domestic fishing industry. For some years, the Atlantic fisheries did prosper. However, pressure on the stocks continued to mount, thanks to the use of bulkier vessels and new technology capable of finding fish like never before. The catches rose, yet cost and market factors drove the four largest groundfish processors almost to bankruptcy in the early 1980s. While efforts were made to shut down or consolidate some plants, local communities fought the closures and resulting loss of jobs, and most of the plants remained open.

On the Atlantic, catches hit a record of more than 1.4 million tonnes in 1988, but in Newfoundland, fishermen warned that Northern cod was getting scarce. By 1989, federal fisheries scientists were calling for a dramatic reduction in cod catches, but cabinet ministers insisted on maintaining higher quotas than the scientists recommended. Enforcement of quotas was also difficult, given the hundreds of group and individual quotas, and fishermen who evaded quotas skewed the catch statistics leading to inaccurate scientific stock assessments.

All this led to that fateful day in 1992 when fisheries minister John Crosbie imposed a moratorium on the Northern cod fishery, followed by closures for other cod, haddock, and other groundfish.

How did it come to that?

Money and subsidies, for one.

The federal government was keen to subsidize the development of a Canadian dragger fleet in order to take advantage of the opportunities created after banning the international draggers from fishing Canadian waters.

Nothing like it, with the possible exception of the fur trade, Cariboo Gold Rush, or Alberta oil and gas exploration, had ever been witnessed in this country. Expectations for profits were unhampered.

Federal politicians were just as giddy. An uncontested cod bounty, because of the economic benefits that were sure to come, would certainly tamp down insistent separatist sentiment that lingered, not far from the surface, ever since Newfoundland joined Confederation in 1949.

As Elizabeth Brubaker summarizes in her Hoover Institute report, successive governments devised dozens of assistance programs for fishermen and fish processors:

> Federal support for the Atlantic fisheries dates at least as far back as 1930, to the formation of the United Maritime Fishermen and the funding of its cooperative processing and marketing efforts. Subsidies increased over the years, generally in response to periodic crises in the industry. They skyrocketed after 1977, as federal and provincial governments alike pushed the fishing industry to take advantage of the new opportunities presented by the extension of Canada's fisheries jurisdiction to 200 miles. In those heady days, provincial loan boards showered fishermen with loans at concessionary interest rates in order to help them buy bigger boats and more sophisticated gear. Between 1977 and 1982, fishermen's indebtedness to the boards increased by 400 per cent to $219.4 million. Provincial officials apparently gave little thought to the likelihood of repayment of their 791 outstanding loans. By 1982, almost half of the loans made to Newfoundland's fishermen were overdue.
>
> Loans provided just one of many kinds of assistance. Governments made cash grants to fishermen and fish processors. They created tax exemptions for fuel and equipment. They acquired equity in insolvent processing companies.

According to Brubaker, Newfoundland raked in the assistance. Between 1981 and 1990, federal and provincial net outlays for Newfoundland's fishery amounted to $2.94 billion, far in excess of the value of the catch.

The most generous subsidies took the form of unemployment insurance (UI) benefits, which, after netting out premiums, amounted to $1.647 billion during that period. During the 1980s, Newfoundland's fishermen relied on UI for an ever-increasing portion of their income. According to the Auditor General's 1997 report, by 1990, they were receiving $1.60 in benefit for every dollar they earned in the fishery, up from ninety-six cents in 1981.

Unemployment insurance (renamed employment insurance in 1997) supported generations of Atlantic fishermen and processors. Introduced in the mid-1950s, the program became progressively more generous; it simultaneously evolved from an insurance program to a permanent income-transfer program. In 1971, six weeks of fishing bought five weeks of benefits; by 1976, eight weeks of fishing bought twenty-seven weeks of benefits. Although eligibility requirements became somewhat more strict in the 1990s, for many years plant workers had to work only ten weeks to qualify for forty-two weeks of benefits.

For many on the Rock, the dole became a way of life. By the early 1980s, UI was widely taken for granted as a natural part of fishers' incomes. In fact, as Brubaker and many others have concluded, a good number of fishermen now consider UI itself, rather than fish, as a resource!

"The way of life of the outports of Newfoundland has been centred on the fishery from the earliest days, but the nature of people's dependency has changed," John Crosbie explained in 1997. "In recent years, their economic survival has depended less on the fish they caught than on their ability to qualify for financial-support programs. Federal unemployment insurance is the lifeblood of rural Newfoundland."

"Lifeblood or not," Brubaker wrote in her report, "it is hard to imagine that anything could have so effectively threatened rural Newfoundland's fishing communities as did their reliance on unemployment insurance. The perverse effects of UI have become tragically clear: It created a dependence that it could not sustain, leaving tens of thousands of fishermen, processors, and their families wondering how they will survive in the coming years."

Unemployment insurance encouraged people to remain in communities lacking any promise of a viable future. More insidiously, the program prompted young people to quit school, leaving many ill-equipped for other jobs when the fishery collapsed and leading to a culture of hyper-aggressive dependency.

Perhaps unemployment insurance's most egregious legacy was its contribution to overcapacity in the fishing industry, which eventually led to the collapse of the groundfish stocks. The program lured workers into an industry that could not support them.

Shortly before his retirement, Newfoundland premier Clyde Wells described the complicity of the provincial governments, federal government, industry, and the workers themselves: "To some degree both governments encouraged the use of the fisheries to create qualification for unemployment insurance. They [the people on the UI system] were induced! They were shown methods by governments as to how to do it! In some cases, fish plants and make-work projects would hire workers for a certain number of weeks and then lay off those workers and hire others, so that they'd all have qualification for unemployment insurance. This was done with the approbation and knowledge of both the federal and provincial governments," Wells told the Harris inquiry in 1998.

As the number of workers in the fishing industry increased, so, thanks to other subsidies, did the capacity of their boats and plants. The 1993 Task Force on Incomes and Adjustment in the Atlantic Fishery summed up the problem: "Too many harvesters use too many boats with too much gear trying to supply too many processing plants by finding and catching too few fish," explained task force chair Richard Cashin.

As early as 1970, a federal cabinet memorandum described a fishing industry that was overcapitalized by a ratio of more than two to one; it estimated that Canada's commercial catch could be harvested by 40 per cent of the boats, half as much gear, and half the number of fishermen.[1]

The Department of the Environment's 1976 Policy for Canada's Commercial Fisheries acknowledged the overcapacity problem in both the catching and processing sectors. But, no sooner had it warned that the long-term viability of the industry depended on getting rid of this structural defect than it started back-pedalling. Any change that occurred would have to be gradual and would have to present acceptable alternative opportunities. The policy stated, "Where adverse social side-effects such as reduced employment opportunities can be kept within acceptable limits, restructuring should proceed. Where damage to the community would outweigh advantages in the short run the changes must be postponed."[2] How the department proposed to reduce capacity without eliminating jobs it did not say.

Pressures came not only from fishermen and fish processors. Between 1974 and 1981, the number of licensed fishermen in Atlantic Canada increased by 45 per cent, to approximately 53,500, while the number of processing facilities increased by 35 per cent, to 700.

Officially sanctioned, torqued by governments, it was easy to tune out the warnings of a few contrarian scientists and academics. The dockside thrummed as an industry geared up overnight. Vessels had to be built, equipped with the latest sonar, and other high-tech gadgets, such as GPS, making cod fishing a form of search-and-destroy in massive steel vessels.

Soon followed an attack on a species that can be best described as military in operation. Thus, began fifteen years of bottom dragging out on the Grand Banks. Trawlers, which are the most destructive of all fishing methods, drag heavy gear behind them along the ocean's bottom, crushing, burying, or uprooting marine species. It has been compared to clear-cutting ancient forests; critical habitat is destroyed and biodiversity is lost. Draggers also routinely dump unwanted fish.

According to the Northwest Atlantic Fisheries Organization (NAFO), the following numbers of Canadian trawlers worked the Grand Banks, fishing cod and other groundfish:

1977 - 147
1980 - 152
1983 - 148
1986 - 152
1989 - 136

This data was particularly difficult to come by in Canada. Repeated requests to DFO were thwarted, disregarded, or referred to other agencies. It was the same story with processing companies such as Fishery Products International and National Sea. There followed a game of referral roulette, which, after several weeks of email and long-distance phone calls, came to an abrupt dead end. Even veteran consultants—never a reticent crew—ran for cover. When it came to the shocking collapse of the Grand Banks cod, and the question of whether the Canadian offshore trawl fleet had done the plundering, few were willing to go on the record. This led one visitor to conclude the mystery of the Grand Banks cod death is far from resolved; that the subject itself raises an acute sense of embarrassment, even shame. Citizens of the Rock resort to a version of the good German defence: no one is to blame because we were all just doing our jobs.

During the 1980s this amounted to as much as 300,000 pounds of fish per trip. As new technologies like rock-hopper trawls were increasingly employed, the damage only accelerated. These gears are able to scour marine ecosystems even in previously unreachable areas. Pockets of formerly intact seafloor were now being decimated.

From a glass tower fronting St. John's waterfront, DFO fishery managers cast a blind eye. In a classic case of groupthink, they didn't want to—or were afraid to—hear about habitat concerns even as they opened new areas and new fisheries to industrial trawling and dragging.

DFO managers joined the queue at dockside celebrations as vessel after vessel, groaning to the gunnels with cod, unloaded their catch at processing plants all over the province, where, waiting only for a shift change, a beer, and a full tank of diesel, they headed back out to the Grand Banks fishing grounds. For a few short years, there was very little reason to imagine anything but a happy ending.

As warnings from nature go, they don't come much starker than the collapse of the Canadian cod fishery in Newfoundland due to overfishing. The cod and thousands of jobs that depended on them disappeared virtually overnight. Now because the cod stocks have failed to recover, seals are being blamed and hunted in record numbers.

By 1990, Canada's offshore dragger and trawl fleet were catching much of the groundfish, peaking at more than 700,000 tonnes. Meanwhile, catches by the inshore fleet—these are the dory men, who for 500 years worked the uncountable bays and inlets of the island—began a significant decline. Two years later the Newfoundland Inshore Fisherman's Association, after a series of desperate pleadings, went to court to try to get Department of Fisheries and Oceans to cut quotas and issue an injunction to stop draggers from fishing the northern cod spawning grounds during spawning season. They lost the case, even as draggers and trawlers bulldozed their way through Grand Banks cod and rearing grounds, literally tearing the habitat to pieces.

This made-in-Canada fleet of midsize trawlers and draggers with huge nets could hoover up massive quantities of fish, quickly processing and deep-freezing the catch, working around the clock in all but the worst weather conditions.

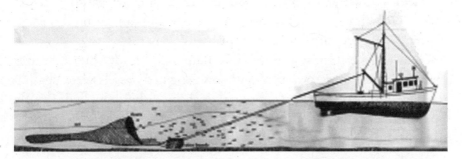

Figure 5.1: Bottom dragging was common on the Grand Banks. Trawlers drag heavy gear behind them along the sea floor, crushing, burying, or uprooting cod and other fish. The most destructive type of industrial fishing, it has been compared to clear-cutting ancient forests.

In an hour they can bring up as much as 200 tonnes of fish, twice as much as a typical Grand Banks schooner would have caught in an entire season.

The cod catch steadily increased to 800,000 tonnes in 1968, but this was the peak of the clearly unsustainable catches. By 1975, the annual catch had fallen by more than 60 per cent. Catches of other fish were also plummeting under the relentless fishing pressure. This forced Canada to extend its fishing limit for foreign vessels from 12 miles to 200 miles from its coast.

Rather than using this rule to reduce fishing pressure on the cod, the Canadian government and fishing industry saw a massive cash bonanza—now exclusively for Canadians. Huge investments and government subsidies poured into the construction of the same destructive factory trawlers so big money could be made from the cod. In the short term, catches rose again and the industry prospered. But beneath the waves, the huge trawl nets were not only scooping up cod and anything in their path, but the heavy gear was plowing up the seabed and destroying the delicate ecosystem. The Grand Banks ecosystem was already on borrowed time.

Factory trawlers systematically emptied the cod from the Grand Banks. Stocks have not recovered since all cod fishing was banned on the Canadian part of the Grand Banks in 1992. But trawlers still fish for cod in international waters off the Grand Banks. As the cod declined, the factory trawlers used powerful sonar and satellite navigation to target the few remaining large shoals of cod, especially during the breeding season when they gather in large numbers.

During the 1980s, cod catches remained steady, but that was because larger, more powerful and sophisticated vessels were chasing the few remaining fish. Traditional inshore fishermen had already noticed their catches declining, but the government preferred to listen to the industrial fishing companies that claimed there was no problem.

Landings and Their Value within Canada's Exclusive Economic Zone (EEZ) from 1950 to 1992 Using Bottom Trawl Catch

Year	Landings	Landed value (2000 constant dollars)
1950	82,687	$50,804,724
1951	91,224	$64,400,639
1952	134,088	$97,033,120
1953	129,543	$93,278,170
1954	190,506	$100,946,263
1955	248,383	$142,854,657
1956	360,851	$216,444,962
1957	358,309	$221,719,730
1958	316,972	$246,831,368
1959	231,676	$189,377,215
1960	440,917	$349,262,601
1961	429,188	$347,080,768
1962	427,188	$381,784,230
1963	462,759	$433,877,864
1964	483,229	$460,889,215
1965	576,248	$582,008,288

(*continued*)

Landings and Their Value within Canada's Exclusive Economic Zone (EEZ) from 1950 to 1992 Using Bottom Trawl Catch (*continued*)

Year	Landings	Landed value (2000 constant dollars)
1966	496,971	$481,208,531
1967	459,605	$410,379,657
1968	494,764	$484,653,472
1969	499,019	$613,148,645
1970	625,971	$774,938,504
1971	730,256	$849,527,404
1972	612,356	$703,614,614
1973	778,885	$886,106,392
1974	580,912	$670,727,509
1975	588,614	$658,729,013
1976	513,956	$652,694,517
1977	408,583	$536,806,919
1978	403,098	$570,659,027
1979	401,225	$514,652,466
1980	388,381	$440,337,236
1981	407,273	$431,304,755
1982	382,420	$418,762,014
1983	330,431	$373,511,720
1984	374,537	$406,030,760
1985	412,709	$475,689,764
1986	472,166	$577,358,298
1987	451,675	$602,894,056
1988	438,728	$573,837,938
1989	432,316	$566,896,996
1990	403,680	$518,031,319
1991	376,679	$500,174,071
1992	356,864	$522,070,621

Data source: Sea Around Us project database (http://www.seaaroundus.org/) and Sumaila et al. (2007)

Scientific warnings in the late 1980s went unheeded because any cut in catches would cause politically unacceptable job losses.

By 1992, the levels of northern cod were the lowest ever measured. The government was forced to close the fishery, throwing 40,000 people out of work and devastating many fishing communities. Despite the ban, stocks have yet to recover and it is uncertain if they will fully recover given the changes wrought on the Grand Banks ecosystem by decades of industrial fishing.

The Department of Fisheries and Oceans recently released a report entitled, *Impacts of Trawl Gears and Scallop Dredges on Benthic Habitats, Populations and Communities.* The report clearly shows the horrific impact of bottom dragging on fish habitats, indicating that ". . . measures to reduce impacts of mobile bottom-contacting gears requires case-specific analysis and planning. There are no universally appropriate fixes." The DFO report adds that, "Recovery time from perturbation by mobile bottom-contacting gears can take from days to centuries, and for physical features and some specialized biogenic features recovery may not be possible."

More than 1,000 marine scientists released a statement calling for swift action to protect imperilled deep-sea coral and sponge ecosystems at the annual meeting of the American Association for the Advancement of Science in San Francisco. They urged the United Nations and other international bodies to establish a moratorium on bottom trawling on the high seas as a precaution "to avoid the very real threat of serious or irreversible damage."

But not only will Canada not sign onto the UN call for a ban on deep-sea bottom dragging, Canada is actually fighting to continue with this indiscriminately destructive practice. Canada is doing damage to an ultra-sensitive ecosystem and turning it into a wasteland, not to mention using weights that gouge trenches out of the sea bottom.

For anyone who has ever seen a video of dragger gear moving across the ocean floor, there is no question that this complete and utter destruction must be stopped at once. It is amazing to note that in the Grand Banks, where cod has become an almost endangered species, bottom trawling is still allowed by the Canadian government. The only action that will work is a complete ban on what amounts to an attack on all ocean life.

Given the vehemence with which Newfoundland fishermen opposed closing off the Hawke Channel, the major breeding ground for the Labrador snow crab fishery, I'm willing to venture that they are also part of the opposition to banning this type of trawl as well.

Having overseen and subsidized the destruction of the Grand Banks fishery, the Canadian government now pays out billions of dollars of taxpayers' money in social security to out-of-work fishermen and communities in Newfoundland. Rather than recognize that it caused the collapse of the ecosystem, it has been busy looking for a new scapegoat.

Because cod stocks have failed to recover, the popular government "common sense" claim is this: it must be because harp seals are eating all the cod and preventing their recovery.

Seals make an expedient target for politicians to blame. The Canadian government increased the seal-hunt quota during the 1990s and in 2003 announced both the permanent closure of the cod fishery and a huge increase in the hunt to 350,000 seals.

The simplistic claim that seals eat too many cod is the same flawed argument (whales are eating too much fish) that whaling nations now use to call for the resumption of commercial whaling. Checking a few simple facts exposes this sham. Cod make up only about 3 per cent of the average harp seal's diet, which constitutes an insignificant amount of cod biomass.

That diet also includes species that eat young cod. There is no science to back the claim that seals are preventing the recovery of the cod. In 1995, ninety-seven scientists signed a petition on the subject: "All scientific efforts to find an effect of seal predation on Canadian groundfish stocks have failed to show any impact. Overfishing remains the only scientifically demonstrated conservation problem related to fish stock collapse."

Away from the cheering throngs, others were watching. What if the renegade scientists, such as Ransom Myers and Carl Walters, were right? What would be the effects if the Canadian offshore trawl fleet continued to hammer away at catch levels that were absurdly optimistic? What would happen to the stocks, the fish that the cod themselves ate, and the complex Grand Banks ecosystem that supported them all?

New data provides conclusive proof as to who decimated the Grand Banks. Ussif Rashid Sumaila, a professor at the Fisheries Centre at the University of British Columbia, has been investigating the cod collapse and his findings represent a fresh way of framing and solving a mystery that has become a national and international disgrace.

A s seen in the data in the table on pages 62–63 provided by Sumaila and his colleagues, there wasn't a huge change in catch after foreigners were barred from fishing in Canadian waters around 1977, the year Canada extended its jurisdiction to 200 miles at sea, effectively evicting the foreign fleet from the Grand Banks.

His research proves that the Canadian offshore fleet continued to pummel the stocks at the same absurdly high rate as had been the case before the "foreign devils" were turfed.

As Sumaila noted, most of the government subsidies were used to pay for the fuel needed to operate out to the 200-mile limit and beyond. Just consider the costs of fuelling such fleets. Deep-sea trawling involves dragging fifteen-tonne weights across the seabed to break up corals and rocks so that fish can be scooped up into vast nets.

Today, noting that Grand Banks cod stocks are 1 per cent of what they were in 1977, Sumaila offers a novel, albeit controversial, solution to bring back the decimated stocks that have plagued the fisheries of Newfoundland and Labrador and countries around the world.

In the spring of 2007, he delivered a presentation to the World Trade Organization in Geneva, calling on governments to eliminate what he considers "bad subsidies." He defines such subsidies as funds that encourage vessels to fish more than they would in the market system, such as subsidies for fuel and boat repairs. He estimates that Canada gives more than $163 million in bad subsidies.

Sumaila is aware his idea raises the ire of fishermen and would require unprecedented global co-operation. But, he maintains the equation is simple. "It's about jobs and fish today versus jobs and fish today and tomorrow," he said.

According to a May 2007 article by reporter Doug Saunders printed in the *Globe and Mail,* after a decade of working in the field of bioeconomics, Sumaila has "found the link between stocks of fish and pools of money, which could save both fish and fishermen. It's a solution that requires the co-operation of every nation, but it could turn fishing back into a worldwide industry that lasts forever and provides food for everyone."

Sumaila came to Canada from Nigeria via Norway, where he earned his doctorate in economics. In 1995, he joined the renowned Fisheries Centre at the University of British Columbia, which had become famous for its director, Daniel Pauly. The acclaimed biologist had pioneered the methods for measuring the world's fish stocks and, it was largely due to his efforts that, in 2006, it was discovered that a third of the world's fisheries had collapsed.

Pauly blew the whistle on overfishing, but nobody had really explained why it had happened. That became Sumaila's role.

"I realized at a certain point that if we can get the roots of the economics linked to the biology," he explained to the *Globe and Mail* reporter, "we can get to the bottom of the problem of decline. I had been doing studies of individual nations and individual species of fish, but when I got to Vancouver, Pauly told me, 'You know, you can keep on doing those little models, but what we really need is an entire global model of the problem.'"

According to Saunders' article, Sumaila's field of bioeconomics had been invented in the 1950s by two Canadians, Scott Gordon and Anthony Scott. "Their model, still being used in the industry, showed that fisheries should be self-sustaining," wrote Saunders. "When stocks get low, or fuel prices get high (boat fuel is the main expense in fishing), then fishing should be less profitable, getting people out of the business and allowing stocks to replenish. Obviously, this wasn't working. But why?"

To get at the answers, Sumaila worked for ten years on creating two huge databases with the help of countless research assistants and travels around the world. One of the databases indicated the price of each type of fish in every fishing nation since 1950, while the other contained the amount of subsidies paid by governments throughout the world to their fishing sector. This data, combined with Pauly's fish-stocks databases, helped Sumaila to discover why the fish were disappearing.

What he learned, according to Saunders, is that "the self-balancing nature of fishing is thrown out of kilter by the widespread government practice of giving fishermen subsidies for boat building and, especially, fuel. This money, which he described as 'bad subsidies,' is exactly equivalent to the scale of overfishing—the subsidies make the difference between a renewable resource and a dying resource."

In addition, Sumaila discovered, one can point the finger at fuel subsidies for the continuation of the most devastating practice in fishing: high seas bottom-trawling.

According to Saunders' research, countries pay $152 million a year in fuel subsidies to trawlers, which accounts for 25 per cent of their income, and their profit is only 10 per cent.

"Without subsidies," Sumaila explained, "the bulk of the world's high seas bottom-trawl fleet [would] operate at a loss, thereby reducing the current threat to . . . fish stocks."

Without so-called bad subsidies (which amount to between $20 and $26 billion a year worldwide), fewer people would be in the fishing business around the world. According to Sumaila, this would actually give the world more fish.

"There is a potential to actually increase the catch if we can agree to reduce the scale in the short term," he told Saunders, "and avoid subsidizing the industry too much in the long term."

While this sounds like the perfect solution—since governments pay taxpayer subsidies to industries for fear of losing the industry and its political support—to achieve it, every country would need to end its fuel subsidies immediately.

Canada, unfortunately ranks as the world's tenth-worst offender, paying out $163 million a year in bad subsidies to its fishermen. As Saunders puts it, "we have, in effect, decided to continue the tragedy of the fish in order to stave off the tragedy of the fishermen"—in the short term.

"We could solve both problems, if we agree to end fuel subsidies in the WTO talks," Saunders continued. "It's an idea that should have appeal to conservative governments. After all, Sumaila has discovered a case where free trade will protect a vital resource, while protectionism and government spending will destroy it."

His ideas haven't exactly resonated in Newfoundland. Earle McCurdy, president of the Fish, Food and Allied Workers Union, said Sumaila's idea is ludicrous and would leave many currently working in the fisheries industry out of work. "These professors got lots of time to sit in the ivory tower and come up with all this stuff," McCurdy said.

Eliminating global subsidies would render these fleets economically unviable. Instead, as fish stocks around shallow continental shelves have declined and collapsed, fishing fleets have gone further afield into deeper waters to catch fish that, in previous decades, were considered too low value to be worthwhile. In essence, the story of the Grand Banks cod.

In the late summer of 2007, the Fisheries Resource Conservation Council (FRCC) issued its latest report on fisheries stock—this time, a stock that is well within the 200-mile limit, and well within the control of DFO and Canadian fish harvesters.

The concern now is lobster, and the federal panel has blunt words about where the fishery is going.

"The fundamental conclusion of the FRCC is that the risks to sustainability are too high in the lobster fishery and that the time has come for the industry to take charge and mitigate these risks. The time of doing nothing has passed."

The made-in-Canada catastrophe of the Grand Banks has shown the world that we can no longer build an economy on the back of a resource. It is an indisputable truth that rebuilding the cod stocks will be long-term and must, by definition, be built on new ideas and a new mindset, not likely to come from governments in St. John's or Ottawa.

Past behaviour proves we cannot trust governments to rebuild the ecology of the Grand Banks cod. The collapse of this stock offers documented proof that politicians and bureaucratic officials were willing to sacrifice a magnificent resource in a direct move to protect jobs and hold on to political power.

Again and again, in the scrabbling history of Newfoundland, politics trumps science.

In 1994, *The Economist* magazine wrote "overfishing and pollution are exhausting the seas. The decline will be reversed only if governments stop subsidies to fishing fleets, police their waters well, and, hardest of all, persuade fishermen to trust them."

Much of this has been known in selected, self-serving bits in Newfoundland for years. But, nearly sixteen years after the Grand Banks cod collapse, industrial overfishing continues rapidly. People point fingers and try to allocate a shameful past to history, as politicians are reluctant to conserve stocks, for fear of reducing fishers' income.

For their part fishers have been unwilling to follow scientists' advice and, on the Rock, heap scorn on DFO, its managers, and fisheries officers. Poaching is considered a birthright by many living in the hardscrabble place.

This is not happening because fishers are foolish. Their behaviour is rational given the signals they face. Like hunters, fishers will try to take what they can when they can, before anyone else catches it. Conserving fish stock makes no sense if the fisher has no right to the offspring. On the contrary, catching more fish today will increase income. Although there will be fewer fish next year, the cost will not be borne by the fisher alone and will be spread over the entire fleet.

For 500 years, fishing has always been this way and any real change will be built on new ideas, trust, and new economics in a global marketplace. Without government subsidies, it would be a very different occupation indeed.

6

Federal Fisheries: A Sad and Ignoble Sputter into Inconsequentiality

"Today, the local salmon canneries aren't industrial sites—they're museums."
—Scott Simpson

What can be said about the Department of Fisheries and Oceans, an agency variously despised, derided, even pitied for its incompetence as it sputters into inconsequentiality? What happened to a once-proud federal agency? Why, on both Canadian coasts, is it now known as the gang that couldn't manage or protect fish—the Grand Banks cod collapse being the most flagrant failing under its watch?

Put simply, Canadians are not getting value for their dollar. In 2007, DFO had a national operating budget of $1.6 billion with 10,500 employees.

It is seen as a tired and cynical crew; a place where overworked minions toil in the regions where aging apparatchiks, some promoted because of their fluency in French, top up pensions in glass towers in downtown Ottawa. It's an agency that, using modern business parlance, hasn't produced value for taxpayers' dollars during the past thirty years.

This was not always the case. In the early 1980s, because I thought I was doing the right thing, I worked in the communications shop at DFO Pacific, drafting—and endlessly redrafting—news releases and speeches for senior officials, often standing by a fax machine late into the evening. The agency had a quasi-military feel, and one soon got used to a parade of middle-aged males padding out to the elevators at the stroke of 4:30 every afternoon. They came by this punctuality honestly: more than a few had been in the Canadian Armed Forces; for service to their country, they had been placed at DFO until their pensions kicked in.

For a "fish flack" like myself there was, at this time, a certain pride about working at an agency so directly plugged into the economic and social activity of the British Columbia salmon industry, then ranked as the sixth most important in the province, employing thousands in the tiny communities that dot the fiords of the B.C. coast.

At DFO, however, every new employee soon learned that, as an agency, we existed to support the commercial fishing industry. Period. Indeed, the Federal Fisheries Act still defines itself as such to this day, despite a series of aborted attempts to modernize itself with verbiage that includes words such as "conservation" and "sustainability." In those days industry ruled and, not surprisingly, executives rotated back and forth between jobs in the private sector and senior posts at DFO, in much the same way retired U.S. senators cross over to work as K Street lobbyists.

When managers from B.C. Packers, Ocean Fisheries and the Canadian Fishing Company and others entered the building, much decorum was in evidence. Industry executives were to be treated with respect that bordered on an unsavoury ceremony.

Swaggering with power, barking out stentorian orders to female typists, this coterie of middle-aged men knew how to rattle the chains of the regional officials simply by picking up the phone and chatting with the minister or his staff. Calls were returned instantly. This clubby insiderism bred friendship; each worked his side of the street, but collectively they had one shared goal: increase salmon allocation for the commercial fishing fleet in order to optimize profit. More fish. More money. Period. The algorithms for these business plans were drafted in small rooms choking with testosterone—and no one dared to open a window.

In those days, decisions about the fishing seasons were made in curious fashion. According to current regional director general Paul Sprout, he would attend the B.C. Packers picnic the weekend prior to the opening and, over drinks and a plate of barbecued salmon, chat with senior managers in this and other private-sector companies, hammering out the details that would define the livelihoods of the salmon fleet.

Sprout relates this story about a kind of old-boys' club in a boilerplate speech, marking the contrast with today: the layers of consultations, the tables of "stakeholders" whose growing list include First Nations, greens, recreational, tourism officials, and a myriad of special interests whose every utterance is transcribed and reviewed, long before the fishing season—or what remains of it for the net fleet—is made public.

Then came a tsunami of change. The fishing industry was shaken from its sleep. The rise of environmentalism and aboriginal rights, on the one hand, was coupled with a growing public awareness that DFO, answering as always to its political masters 3,000 miles away in Ottawa's Mind Control Central, simply did not have the interest or the capability to protect Pacific salmon stocks.

In the mid-1980s, in a scenario that simply could not happen today, a trio of ambitious young reporters, all working for the *Vancouver Sun*, competed with each other to investigate and file important stories about salmon. They bombarded the city desk of a paper that, in those days, covered fisheries like a blanket because, like forestry and mining, they were the backbone of the provincial economy.

Mark Hume, Terry Glavin, and Scott Simpson had been paying attention. They had ferreted out the documents to prove the salmon stocks were threatened, and they knew this information would shock the public because salmon were such an icon in the economy and a way of life to aboriginal and non-aboriginal peoples who lived in close-knit coastal communities.

Each reporter, in his own way, became a quasi expert when it came to issues such as salmon allocation, habitat destruction, and stream enhancement. In small boats and from float planes, they visited ragged landscapes and deep fiords to see first-hand the follies of big-budget hatcheries and the aftermath of oil spills. Learning that fishing stripped of its mythology is about catching and killing to make a living and pay the rent, they spelled out how DFO tried and failed to juggle the deadly warfare between commercial, aboriginal, and recreational interests—which led directly to overfishing. Hume, Glavin, and Simpson correctly identified the struggle, now being played out so disastrously, that would threaten and decimate stock of wild salmon.

The politics of fish were, and remain, fierce and take-no-prisoner. In story after page-one story, the general public was getting a primer, and the news was often bad: botched science, muddled management, questionable priorities, trumped, it always seemed, by politicians in Ottawa.

Picking and probing beneath the official story, the reporters sometimes embarrassed senior staff into telling the truth about salmon. Those truths were of a common theme: wild salmon stocks are being decimated, some to near extinction, and DFO, despite endless spin and study, was powerless to stop the killing.

With the energy and innocence of youth, the *Sun* trio probed beneath the hallowed and, as it turned out, often sanctimonious and self-serving mythology of the lone fishermen set against the terrible forces of nature. They were writing a new narrative—one that began with the assumption that catching and killing fish was for the commercial fleet a way to make a living, raise the kids, and pay the bills. Other articles revealed human greed and a stupidity borne of a willful ignorance. There followed stories of "fish hogs" near Bowser on Vancouver Island, how sport fishermen every day of the year trolled the waters of the Georgia Strait to "limit out" and returned, each day, with four or five chinook to pack into freezers inside motorhomes with U.S. licence plates parked at a nearby marina.

There were myriad "bycatch" articles that detailed the slaughter and waste wherein other species were thrown dead back into the sea, because a particular "gear type" was targeting another species. There were far too many stories about poisoned rivers and, sadly, how cities—to this day—pipe tonnes of excrement into the aquatic ecosystems used by six species of salmon and dozens of other stocks. A litany of horror; a sea of slaughter; a page-one sea of ink is what the trio wrote about.

The right place at the right time. In 1982, University of British Columbia resource economist Peter Pearse released *Turning the Tide*, his Royal Commission on Canada's Pacific fisheries. Provocative and encyclopedic, it delved into the state of Pacific fisheries. As a result, it shone the spotlight on DFO, an agency that, with a confused mandate, suspect intellectual leadership, ever and always, so exactly reduced to the machinations of political masters in Ottawa, was simply unable to make the timely and appropriate management decisions needed to protect and conserve increasingly threatened wild salmon stocks.

This, too, was covered by the *Vancouver Sun* and led to a bigger discussion: the core attitudes and mindset of fishing—which, indeed, had shaped the very idea of fishing for 500 years—became a subject for public discourse. Pearse also dared to frame the fisheries debate in economic terms, urging the federal government to make sweeping changes that would see the fishermen themselves play a much larger role in the management and conservation of the stocks. If insiders had identified these issues for some time, few were ready for Pearse's blockbuster: he accurately predicted that First Nations would, over time, win legal access to more salmon. That whether non-aboriginal fishermen liked it or not, the aboriginal fishermen would have to be accommodated and included in every aspect of Pacific fisheries.

(Pearse was prophetic. In 2004, he revisited this subject in a report co-authored by Don McRae, dean of the Ottawa Law School. The two wrote that once aboriginal treaties were ratified in the province of British Columbia, aboriginal people would be allocated 33 per cent of all Pacific sockeye salmon.)

Every issue, especially in the media, has its day. Life intervenes. People move on. After four years in the flack shop, I quit the DFO for a freelance career. A farewell pint at the Jolly Taxpayer pub with a coterie of DFO colleagues, surprised that I would willingly pull the plug on a government job, provided fresh insight into their thinking: "How could you leave, Rose? Think about your pension." This point was driven home by an economist who held up a number he had scribbled on his bar napkin. If he stayed in service another twenty-five years, he would receive an indexed pension worth more than $1 million.

The *Vancouver Sun*, still the dominant broadsheet daily on the West Coast, has changed ownership several times, even as circulation has declined. Reflecting an increasingly urban population, its focus grew more metropolitan in scope as resource issues such as fisheries were punted off the front page. Today, oil and gas, investment banking, and high-tech stories, fill its pages. Fading into the sunset, the $161-million wild fishing industry is hardly a nibble in the *Sun's* diet.

Terry Glavin quit the paper in 1993 to write a series of books, including *Waiting for the Macaws*. As a consultant he worked with the David Suzuki Foundation and Greenpeace, and became a founding member of the Pacific Fisheries Resource Conservation Council. "I have been fortunate to have had many environmental organizations subsidize my fascination with things that move beneath the surface of the water," he says.

He also co-authored a report that led to management changes to protect dwindling stocks of groundfish in the Strait of Georgia. Like Emile Zola, Glavin wrote with a finger-pointing style, backed up by encyclopedic facts that he gleaned by himself to pin his victims against the wall. Crusader, advocate, and free-thinker, he deserves every credit for introducing these subjects to a wider public.

Mark Hume, now a Vancouver columnist at the *Globe and Mail*, continues to write about salmon and related resource issues. While lamenting the demise of DFO—"back then it was the lead federal agency on the coast; today it hardly registers"—he says he still gets emails from *Globe* readers when he files a story on salmon.

"People care deeply about salmon and they know that if B.C. loses its wild stocks, it would be a terrible tragedy. DFO managers have never seemed to get this . . . it's as if they have translated salmon into statistical units and have no more feel for the declines in runs than an accountant has for the deletion of numbers. The editors at the *Sun* never appreciated the remarkable coverage they were getting from a handful of reporters who thought that DFO and the mismanagement of the resource was one of the biggest stories in the province," Hume says today.

Senior editors at the *Sun* were even heard to complain that there were "too many stories about fish and Indians," says Hume. He went on to write three books about the wild salmon rivers of B.C.

Scott Simpson, now a business reporter at the *Sun*, still writes about fish but has to do so "off the side of my desk," as editors increasingly assign him stories about mining, petroleum, and high-tech. When he does, he says, "DFO doesn't always call me back. I have to hunt them down. I guess they have to run everything by Mind Control Central."

In a 1988 interview with Simpson, Paul Sprout conceded that the fisheries department's own salmon policies were seriously damaging a number of wild salmon and steelhead runs into the Skeena River system, a sprawling network of salmon streams emptying into the Pacific Ocean near Prince Rupert on British Columbia's north coast.

In the early 1970s, the department, through its salmon enhancement program, had successfully created a thriving commercial fishery based in Prince Rupert by dumping fertilizer into Babine Lake, setting off a population explosion among a formerly sub-par strain of sockeye. Suddenly nourished in a way that nature could never have sustained, the annual returns of Babine sockeye leapt into the millions of fish, making commercial fishermen rich—but touching off an environmental and social disaster that continues to this day.

Each summer a gauntlet of commercial gill nets greets the sockeye as they return from the ocean to commence their spawning migration up the Skeena, then off into the Babine River and the lake that feeds it. Few of the fish make it through to spawn—but the fertilization program works so well at boosting the amount of food in the lake that the survival rate among sockeye offspring, or fry, is so high that it seems no amount of fishing effort can wipe them out.

The same cannot be said for other species of salmon and steelhead headed for the Babine, and more than a dozen other major streams that feed into the Skeena. They encounter the same gauntlet, and their populations have suffered greatly over the past thirty years as they are trapped in the same nets as those sockeye.

Those other runs have declined 75 per cent or more in many instances, to the detriment of aboriginal food fishers, tourism business operators, and sport fishers—and, of course, the fish themselves, which might recover if the Babine enhancement project were abandoned.

"Based on the information we have now, we probably would have directed the enhancement in a different way," Sprout told Simpson in that 1988 interview.

Regrets, however, are no catalyst for action by the department. The interest of the commercial fishery continues to prevail.

On June 25, 2007, nearly twenty years later, a series of internal department memos obtained by conservation groups through Canada's Access to Information laws gave some hints of the disgust that fisheries managers themselves feel about the political priorities that trump the best interests of the fish. Steve Cox-Rogers, a federal stock assessment biologist, in a frank assessment of his department's motives, told a B.C. government colleague that DFO "caved" under pressure from Prince Rupert mayor Herb Ponds and kept open the Babine sockeye fishery in 2006, despite grave concerns that a sustained opening could prove disastrous for other populations of salmon and steelhead that would be intercepted.

"Our mayor flew to Vancouver to get DFO to provide more fishing time and so we ended up fishing a few more days," Cox-Rogers wrote in an October 11, 2006 memo.

Even more damning, Cox-Rogers noted that commercial fishermen were using none of the conservation methods they had grudgingly agreed to employ after two very public and bitter decades of debate in which conservation groups and the provincial government fought to persuade the gillnetters to accept responsibility for the damage their fishery continues to cause.

"In fact, all of the fishermen I spoke to expressed little desire to participate in reviving steelhead or coho and were just throwing them back dead or alive as they hit the boat," Cox-Rogers wrote.

Scott Simpson, fifty-two, says the number of conservation and environmental groups raising the alarm about the decline of wild salmon has increased exponentially in the past fifteen years, but their laments for the fish are playing to an ever-shrinking audience. A twenty-six-year veteran of the *Vancouver Sun*'s city desk, he's contributed hundreds of volunteer hours to fisheries enhancement programs on a variety of streams around B.C.'s south coast. Among his various journalism awards, he is the only two-time winner of the Art Downs Award for environmental writing from the British Columbia Wildlife Federation. The first award from the federation was for a series of stories about a proposal to put a gravel mine on the Upper Pitt River. The stories were credited by environmental groups for the B.C. government's decision to revoke the mine proponent's permit. Simpson notes, "The proponent also tagged me as the catalyst for that decision, although I would say I was blamed, rather than credited, in that circumstance."

A soft-spoken man with a wry sense of humour, Simpson has lost all patience with the DFO, incensed that the agency is too weak and too corrupt to do its duty.

Today, the local salmon canneries aren't industrial sites—they're museums—and if commercial fishermen actually get to go out on the water, it's kind of newsworthy because it happens so infrequently. But the reality is, a story about fish has to compete for space in the paper with any other story on a given day. And, let's be honest. In this province, in this century, the Fraser River generates more debate in its role as a transportation bottleneck for commuters who want more bridges over it, than debate about how to protect the salmon that use it.

"These days, for better or for worse, more British Columbians get their salmon from a farm than a boat."

Simpson also points to changing public attitudes as the collapse of local coho runs led to sport fishers abandoning their once-loved pursuit for good. He recalls that the *Sun* itself used to sponsor an annual salmon fishing derby that attracted thousands of anglers to Vancouver. The derby has long since been replaced by The Sun Run, a family-oriented ten-kilometre run.

"I think people in general have a sort of nominal concern about salmon, but it's not on a level that would translate into votes for any politician who decided to champion their welfare," says Simpson.

"Even the fisheries department seems to be less involved," he adds noting that groups like Environment Canada, the David Suzuki Foundation, the Watershed Watch Salmon Society, the Pacific Fisheries

Resource Conservation Council, and even the provincial and municipal governments to some limited extent, are stepping in to take on the work of salmon conservation.

"Some of these groups have moved in to fill the vacuum left by the downsizing and demoralization of the fisheries department. About the only thing the department gets to do anymore is take all the blame when something goes wrong or when some group decides it's not getting its fair share of what's available."

But Simpson hardly feels sorry for the DFO. In 2007, he explains, only 1.3 million Fraser River sockeye returned to a system that once saw more than 20 million fish less than two decades ago. That year, "a lot of people got nothing," he says.

"The prevailing wisdom is that climate change is adversely affecting ocean temperature and food supply, and making freshwater habitat less hospitable. Certainly it wasn't overfishing."

Interestingly, pink salmon seem to be suffering as well but, because they're regarded as a low-grade species and of no real commercial value, the DFO has "more or less been ignoring them."

"Nonetheless, when the Pacific Salmon Commission reported in September that pinks were returning at only about half their projected abundance, I thought it was time to take a closer look," says Simpson. The commission also noted that the pink salmon were arriving about nine days earlier than usual.

"That set off alarms in my head because that's the same pattern of behaviour exhibited by Fraser sockeye in many years since the early 1990s, and the effect of that behaviour has been catastrophic for those fish. They show up early, and then their internal time clocks seem to jam."

As Simpson explains, fish that should be spending that time in the ocean "bulking up" for the migration and spawning period ahead are instead milling around the mouth of the river, getting weaker and falling prey to infectious diseases that are endemic to salmon migrating into fresh water. The commission had reported that this was the third consecutive year in which pink salmon were exhibiting that unusual behaviour.

"Even if pinks don't support a commercial fishery, they are a significant part of the Fraser ecosystem," says Simpson. "Even in a bad year, you might see 5 or 10 million coming back. When they spawn, their carcasses provide a huge bump of nutrients into the lower Fraser. It's the same sort of idea the fisheries department has used to such controversial effect on the Babine—all those dead pinks act as a sort of fertilizer at the bottom of the Fraser's food chain, making a rich broth that ultimately nourishes all the young salmon and all the other species that live there and need food to thrive."

Yet when Simpson called the fisheries department—its Pacific region has an annual budget of $250 million and 2,200 employees—to find someone to interview about the situation, he got nowhere. "As best as I could determine, there is nobody at the department who felt knowledgeable enough about the situation to talk to me.

This was disappointing because I'd just written a story based on a report from the Pacific Fisheries Resource Conservation Council, which was calling on DFO to undertake a more ecosystem approach to managing the Fraser system on the premise that the department's fragmented approach, predicated on the interests of user groups rather than the needs of the whole Fraser ecosystem, was failing the river and its fish, and the province of British Columbia.

"After that report came out, a senior biologist with the department assured me that an ecosystem approach was exactly what the department was preparing to take. If that's the case, I wondered, why wasn't anybody watching what was happening with those pink salmon?"

While researching this book, the author requested a series of interviews with DFO staff—from senior officials to communications staff—on both coasts and in Ottawa's "Mind Control Central." All requests were initially ignored. Indeed, the search for even the most basic of facts relating to the Grand Banks cod and Pacific salmon became a frustrating and, in the end, absurd experience. Most information came through the back door, gleaned from unofficial sources—most people spoke only on condition of anonymity; they were clearly afraid they would lose their jobs and pensions—and from academics and experts working in other agencies, most outside Newfoundland. The interview below is the one exception. Because

Deputy Minister David Bevan was the only senior official willing to go on the record, it has been taped and transcribed below:

Author: Who, in your department's view was primarily to blame for the collapse of the Grand Banks cod?

David Bevan: Okay. Well, there's who's responsible for the collapse of the Grand Banks cod, and I think that it's been reviewed a number of times, certainly there has been a lot of soul-searching as to what happened. It was clear that virtually everybody involved had some role as well as a change in the environment. So we have seen that the DFO was perhaps optimistic on the amount of biomass, that the catch was set too high in retrospect, that the changing ecosystem wasn't factored into the fish management, that industry was engaged in unsustainable practices and under-reporting their catch and high-grading, etc., and that there was also fishing by the foreign fleet on parts of the stock outside the 200-mile limit. So, all of those factors, coupled with the significant change in the ecosystem at the time, and reduction of productivity are all parts of the problem, and that was certainly verified in the Cashin report and the report from the FRCC that noted all of those factors. I think it's dangerous to start to say that, okay, one body is accountable because that absolves everybody else from their role in it. And that's still going on to a certain extent, where people take a simplistic view. It was either DFO or the foreign fleet but they're saying, "I didn't have anything to do with it. My role doesn't have to change because I didn't cause it." But that's a real problem because I think that everybody engaged has to look at this thing and we have to figure out that perhaps we weren't cautious enough. We didn't factor in the ecosystem in our management and we have to make those changes in the future and that we are making those changes in how we manage fisheries so we can avoid this from ever happening again in other species.

***Author*:** Did the DFO ever formally take responsibility for its role in the collapse? Become accountable? Apologize?

DB: Well, I think we've changed how we've managed fisheries. That indicates that we recognize that what we'd done in the 1980s and 1990s had to change and wasn't sustainable. So, we changed that. Our big problem, as I said again, is that kind of approach absolves everybody else. Everybody has to change: the fishermen have to change, they can't engage in those wasteful practices that contributed to it, we can't be pushing either the provinces or fishing industry or . . . Communities pushed us very hard to keep the taxes high as they possibly could and it was impossible when the red flag finally did go up to make those adjustments. So, I think for one party to say, "Yeah, we did it," would really be a disservice to moving us forward in a proper way. We've reviewed it, looked at the problems, said okay, we can't be setting tax that high, we can't use commercial data to ascertain how high the stock population is because it was either fraudulent in some cases or it maintained high catch effort right up to the end of the population. So all of those things we've changed, we've recognized it, but I think everybody else has to look in the mirror and say, "Okay, I contributed to this too."

***Author*:** Is Pacific salmon "going the way of the cod"?

DB: I think there is some obvious concern about Pacific salmon. There's a huge difference, mind you. And that is that this time around we certainly recognize that the oceanographic conditions do contribute to problems. So we've seen a lot of attention paid to marine survival of salmon. Studies on the high seas that we do in collaboration with other countries have demonstrated that there has been an oceanographic shift, and we've seen, for example, this year the Fraser River, no fisheries because of lower returns. The big difference here though is that "no fisheries" is what has happened, not "no fish." We have shut down the fisheries in order to allow escape to continue notwithstanding the difficult oceanographic conditions, and people equate, I think, "no fisheries" to "no fish" in the British Columbia context. And we are concerned about other ranges because of the changing conditions and the Fraser River watershed under so much stress with the infestations causing lots of trees to be removed, etc. That's our future concern, it's a concern now, we have to be very cautious. As early as the early 1990s we used to do what was called "stock aggregate management." We wouldn't manage to the weaker stocks; we'd manage to the strength of the larger components of the stocks. Now we don't do that and you can see that for Skeena River we had a very low harvest rate on Steelhead in order to preserve it and that had impacts on the commercial fishery. And the same thing in the Fraser River—we managed those fisheries to protect weaker stocks now.

Author: After the collapse, why did DFO reduce its number of scientists? Wouldn't that be the time to muscle up?

DB: Well, we had a general problem with budgets obviously in those days, and they cut 30 to 40 per cent of the department's budget and without cutting services. And it just didn't hit science, it hit us all.

Author: Since the collapse in 1992, what has been the total cost of social assistance and retraining programs?

DB: We have a number. We can give you some information on the breakdowns. About $3.9 billion in direct costs by the federal government for things such as licence retirements, early retirements, vessel support, income replacement, training, counselling, admin, regional and economic development, and so on.

Author: This is federal money?

DB: That's federal money going into the hands of individuals and communities in Newfoundland and Labrador. Clearly, there's been a lot of people move out and that has a big impact on the economy. We don't have the stats figure . . .

Author: How would you describe DFO's cod Grand Banks fishing regime now?

DB: [It] supports the fishery and there's also the Gulf cod. So, there's 21,000 tons or so of cod available for catch by Newfoundland and Labrador people. So, that's why you see lots of cod around. But there's a very limited fishery on what's called Northern Cod.

Author: And it's got the recreational and the food fish, right?

DB: That's correct. And there's a small inshore commercial fishery that helps provide information about distribution and the population makeup, and that's based on each fisherman having the possibility of using gill nets or hooks to fish in their local area and they get around 2500 pounds per vessel.

Author: The management regime seems complicated. One expert calls it a hodge-podge. How would you describe it?

DB: It is complicated, but people sort of look at their neighbours and try to uh . . . people are looking at how their neighbours are behaving. We have a lot of fishery officers—100 fishery officers or so in Newfoundland and Labrador. They do monitor the fishery and it is complex but because it is so restrictive it keeps people fairly close to shore and it is easy to monitor from vessels operated by fishery officers.

Author: Is poaching a serious problem in Newfoundland?

DB: Well, that's one thing we have to be very cognizant about. That's why we have across the country 700 fisheries officers. They're there to ensure compliance and we have violations to be protected in all fisheries, including those in British Columbia, and take action to deter illegal action. But we have about two billion dollars' worth of landed value. That much of it is readily accessible close to shore and you can appreciate that people would like to get a piece of that.

Author: You are trying to modernize the Fisheries Act. What's the status now?

DB: The minister will be considering what happens next. It died on the order paper with the proroguing of parliament. There has to be some consideration as to whether or not they're going to reintroduce it in this session. That's up to the government at this time. A lot of work went into it and actually it was informed by the experiences with respect to Northern cod. Oh there's references to a cautionary approach, references to ecosystem-based management and so on and so forth in the new Act.

Author: Would it be fair to say the collapse was one of the reasons for the proposed new Act?

DB: I think it is just time. It was almost 140 years old. The current Act provides all power for decisions as to who fishes, where they fish, how much they fish and so on. Every detail is left to the minister's absolute discretion. No guidance to the minister and no process for transparency and so on, so that kind of governance can lead to problems and it was time to modernize it.

Author: Thank you for your time.
DB: You're welcome.

On November, 29, 2007, as Bill C-32, the revised Act, was tabled in the House of Commons, its passage by no means certain. Meanwhile, on both coasts, Canada's imperiled fisheries remained governed by a document that was more than 140 years old.

7

Death in the Outports:
A Town Called Fortune

And I spent my whole life, out there on the sea
Some government bastard now takes it from me
It's not just the fish, they've taken my pride
I feel so ashamed that I just want to hide.
—From *Fisherman's Lament* by Great Big Sea

Utter devastation should never look so lovely.

By late summer, a postcard: the town of Fortune, near the southern tip of Newfoundland's Burin Peninsula, is bathed in a translucent light that would please a painter. Sunlight slanting in from the south, reflected off the sea, picks out a row of tidy, salt-sprayed saltboxes perched on a spit of granite here at the edge of the Grand Banks.

Up top, the widow's walk of a dowdy bed-and-breakfast affords an unobstructed 180-degree view of this great big sea, the fabled and frightening body of water that has, in almost every way, so absolutely defined every aspect of life on this rock-ribbed island for 500 years.

The moratorium on cod fishing, the lifeblood of these communities for 500 years, wiped out 40,000 jobs, changing forever the history and landscape of thousands of Newfoundland outport communities, leaving only a bitter history and the postcard sensibilities of tourist brochures.

Fortune has never been the same. The town has been dying a slow and persistent death since 1992.

In the summer of 2007, after examining sheaves of technical documents and data, I travelled to Newfoundland to investigate the facts of the cod collapse. In St. John's and in the outports and bays, I interviewed hundreds of people who shared their stories. In my transcription of the interviews, my transcription of a Gaelic-infused vernacular, I find documented proof, buried deep beneath the official story, that explains why so many have had so little to say about one of the world's worst ecological catastrophes.

In modern times, no other country has so absolutely destroyed a sustainable resource once considered the eighth wonder of the world. No other country—Brazil and its rainforests one possible exception—has practised killing with such rapacity, with such greed, such monumental stupidity.

The truth of the cod collapse is an inconvenient one indeed, especially for a country that so incessantly hectors others with a made-in-Canada morality; particularly when it comes to conservation and sustainability. Today, playing the blame game, there are those who affect a philosophical view, that everyone is guilty of killing the Grand Banks cod, and so, quote the parable in the Bible in which no one dares throw the first stone.

Wrong. Categorically, indisputably wrong. Surely, the first step in any authentic remedial action is to accept responsibility. Because only then can we set a course of action for the regeneration of the Grand Banks cod. In sixteen years, no one has stepped forward to take responsibility for this sea of slaughter. Not politicians, nor scientists, corporations, managers, or fishermen themselves. It has been a conspiracy of silence.

In the summer of 2007, a visitor made his way down the Burin Peninsula and came to understand that cod was directly linked to every historic event and every building in every village he visited. Here can be found the wild, boreal seascapes for which the island is justly noted, as well as historic villages where one could start to unravel the various strands of Newfoundland's colourful history.

The pretty towns of Burin are gradually seeking more tourism to offset the collapse of the fishing industry. In towns such as Grand Bank and Fortune, a series of high-end bed-and-breakfasts, often heritage buildings with commanding views of the Grand Banks themselves, tower over tidy homes of white clapboard and wood shingles, with handsome old churches tucked on small lots along narrow lanes, many edged with picket fences. Each village has a dollhouse scale to it, and is reminiscent of both Ireland and New England. Yet it remains distinctly Newfoundland, with its squarish, no-nonsense white homes trimmed with tropical colours.

A visitor will succumb to a peaceful daydream when wandering the winding streets—more lanes, really, of a half-empty town—visiting cemeteries and sitting on a bluff looking out at the Atlantic Ocean. One can begin to imagine a past of cod drying in front of many homes on wooden racks. The sour-sweet aroma wafted across the road, suggesting maybe, that one day, the cod will return.

Once each outport had its wharves and warehouses, factories producing cod-liver oil, and a payroll including blacksmiths, boat builders, and storekeepers. When the men, in schooners and dories, were working the Grand Banks, the women and children would climb the stairs to the widow's walk to see if a black flag, signifying death at sea as was very often the case, was flying on the schooner mast. If all was well, and the catch had been good, the vessels would be unloaded in a disharmony of shrieking seagulls, the village would share in the celebration and, on Sunday morning, handsome old churches would swell with Anglican and Catholic hymns.

But King Cod is dead now. Village life is gone forever.

D own in Fortune, past the shuttered fish plant, past the hulks of a fleet rusting idle at the dock, a clutch of tourists lurches off the catamaran and straggles back onto land after the twenty-five-kilometre run back from St. Pierre and Miquelon. One does not have to be bilingual to understand the crossing has been less than genteel. A broken two-metre chop with Force five winds adds up to a wicked, running sea.

The sole remnant of France's once-extensive possessions in North America, the Atlantic islands of St. Pierre and Miquelon lie off the Canadian island of Newfoundland. The ferry run is one of the few reasons that people come to Fortune anymore, and then only in summer.

Since the cod collapse, this once proud outport has been gutted and filleted like the fish that once sustained it. More than 2,000 people lived and worked in what used to be a thriving community, its residents either fishing or processing Grand Banks cod. Now, fewer than half remain, and they are the old, the pensioned, the resolute. Some practise an obstinate denial. Most live off the mathematical formulas of complicated federal government schemes of social assistance provided since the 1992 cod collapse.

Walk the main street of Fortune to see the faces: many are old, worried, and lost in distraction. Sixteen years after the moratorium, they have stopped believing King Cod will ever return. Despite brave talk—the latest is oil royalties from Hebron—and a patented made-on-the-Rock optimism that stretches long into a hardscrabble past. In fact, there is compelling evidence that the billions of resource-based profits exported every year far exceed the net income received by federal transfer payments.

Ironically, there is a new kind of economic activity here today: vulture capitalism. As the old people die, and the young leave for the oilfields of Alberta, homes here are going up for sale. Up and down the rugged coast of Newfoundland, in centuries-old fishing villages like this one, foreigners are taking advantage of a struggling regional economy to buy seaside summer homes. Not long ago, the number of foreigners in Fortune could have been counted on one hand. Today, the summer people arrive each June and July from Tennessee, Ohio, Washington, California, New York, and Massachusetts. And, from nearby St. Pierre, of course.

This same scene is repeated down the road in nearby Grand Bank and in countless postcard pretty outports right around the province.

In Grand Bank, a handful of high-end bed-and-breakfasts regularly book Americans and Europeans, far more today than just two decades ago. The tourists come to mellow out from the frenzy of big city life, to nose about the history of this fabled place now allocated to memory. Several times each summer a big sailboat may arrive from Boston, but this is rare. Only a skilled skipper can navigate the unforgiving waters of the Grand Banks, a forbidding body of water that has, over the centuries, taken the lives of so many dorymen. By late September, the visitors have gone and cold mists and terrifying winds batter the coastline. As near constant rain, fog, and drizzle drape over Fortune, many bed-and-breakfasts, especially those without insulation, close up for winter.

Even with the Canadian dollar hitting a thirty-year high against the U.S. dollar, the tiny white cottages can be found for as little as $10,000, the price of a used car. Realtors claim the cheapest places to buy recreational properties tend to be the places that are the farthest from major Canadian cities.

Not including a six-hour ferry ride, the drive between St. John's, the capital of Newfoundland, and Montreal, the closest major Canadian city, is 1,600 miles. Direct flights from both the United States and London, however, have increased in recent years and that has helped to attract more Americans and Europeans.

Here in Fortune, the influx is making some locals uneasy. Some say the Americans have built enormous homes utterly out of sync with the area's architecture and history. The jump in demand for local real estate by foreigners as well as among mainland Canadians comes as the province of Newfoundland continues to endure a prolonged out-migration.

Since the 1992 cod moratorium, the population of Canada has risen by 16 per cent, while the population of Newfoundland has fallen by 11 per cent. The majority of the 61,000 people who left the province were from rural fishing villages like Fortune, where populations have routinely shrunk up to half. Before the moratorium, Fortune was home to 2,000 people. Today, its population is less than 1,000. Many headed west to work in the oil, gas, and mining industries.

In 2006, more people died in Newfoundland and Labrador than were born. The population peaked fifteen years ago at 580,000 and dropped 70,000 by 2006. In 1971, half the population was under twenty-one and the median age was the lowest in all of Canada. Today, at forty-two, it is the highest. In 1971, this province had 163,000 kids in schools. This year, that figure has fallen to 70,000 and is falling still.

When a visitor sees the empty schools and abandoned sports fields, he begins to understand the pervasive sense of despair that, like the Atlantic mists and fog, cloak the island.

In his editorial in a national newspaper Memorial University political science professor Michael Temelini wrote that "Canadians simply do not know, or care, that this 500-year-old civilization is disappearing. We are witnessing an entire generation without hope, enthusiasm or meaningful, steady employment."

Mere steps from the ferry terminal, in the shadow of the shuttered Fishery Products International fishing plant, Charlie's Bar serves as a community centre for the locals. Here in late summer, tourists are introduced to a fast-talking Newfoundland accent that falls somewhere between Dublin and the East End of London. Talk is friendly and inclusive. But, once the summer people have gone home, the regulars reflect on the death of their single-industry town. The tone is modal and minor key; an Irish lament.

Locals stop by and fall into easy conversations, which embrace scores of people, their illnesses, marriages, relationships, and deaths. Here, in Fortune, it is still possible to know almost every family. As they speak, extensive networks of relationships spring to life, conjured back from the dead and mingled with the living in the afterlife of the unforgotten.

Owner Frank Fizzard is a former schoolteacher, a man of clever asides and a certain sophistication. Fate, he ruefully acknowledges, has cast him a role in a tragedy. He must bear witness to the death of Fortune. Like almost everyone here, he has his theories about who killed the Grand Banks cod and its effects on Fortune and the rest of Newfoundland. Attuned to every nuance of outport life, he hints— but won't quite say—that foreign overfishing, bunk science, and bad management—from Ottawa, of course—led to the cod collapse. He really doesn't want to go on the record. But, he will admit that he knows more than a few people who consider Confederation a fraud and disaster: we should have joined up with the Americans instead.

(This sentiment was a staple of outport life. Resting two nights in a tidy bed-and-breakfast near Marystown, I was charmed by the attention and huge plates of food proffered by the hostess, a widow and grandmother of six. Rattling around the place, she doted on her guests, mixing shards of local history with the story of her ancestry. She may have been lonely, too: all her children had left Newfoundland to look for work—in Boston, Halifax, Toronto, and Calgary. But she still made bread every morning, working the flour with what can only have been the largest pair of hands on the Burin Peninsula. Friendly with a lively and inclusive sense of humour, she shared, over a glass of sherry one evening, another side of her personality: a hatred for all things Ottawa—its fisheries department, its scientists, its managers, and what she called fish cops. Ottawa alone, she insisted, as her voice hardened, had brought suffering to rural Newfoundland. Policies dreamed up in Ottawa were directly responsible for the poverty, high unemployment, and every other aspect of social dislocation and bewildered malaise that so defines outport life today. "We should have never joined Canada, in the first place," she said. "The Americans wanted us. We should have gone to the south.")

Like almost everyone else, Fizzard's children left the province to find a better life. A life with scope and promise. A life beyond the "six-and-two" of state-sanctioned forms of welfare and make-work programs in a province with the highest rates of employment insurance.

If, as befits the owner of a pub, Frank has perfected a relaxed presentation to the world, his deadpan observations are delivered with a kind of weary cynicism, reflecting painful truths. He and his wife will soon join their daughter in her new home in Alberta.

Here in Fortune, people have little discretionary income. "I have to keep my prices down," he says. "Bottled beer is three dollars here. If I moved it up to $3.50, they just wouldn't come anymore."

He points to Luke, the local fisherman, who chuckles, a practised routine. But today, they have an audience. A former cod fisherman, Luke, is a regular at Charlie's. He now fishes lobster from a rusting thirty-nine-foot trawler. Like many here, he has a story to tell. It goes back in time, down the centuries of schooners and dories when cod was king and the most dangerous job in the world was one of the most lucrative and most prestigious. To a visitor's ear, the accent evokes the speech of settlers from England and Ireland who were drawn here centuries ago by the cod.

Proud, fiercely independent, and brave beyond all meaning of the word, they skippered the Grand Banks schooners, now showcased in a nearby museum. During this great and noble age, the men fished for cod and the women salted and dried them on the flakes just down the road, and across the province.

Along the waterfront were the fish-merchant warehouses, where the salt cod was brought ashore, dried and stored, and later stacked in the holds of ships bound for England, Europe, and the West Indies, where it was consumed at ten times the rate per person that it was in Newfoundland. Barge cranes moved back and forth all day, loading, unloading stacks of salt cod the size of houses. The technology of preserving fish had not changed in 500 years. Soak it in brine until its every fibre is so salt-saturated it will be safe from rot for years, and then dry it in the open air.

Salt cod would lie drying everywhere within several hundred feet of Fortune's harbour. It lay on the rocks in backyards and on elevated fish-flakes near the water; cod split and cured in brine and set out to dry. When it rained, everyone rushed out to turn the cod over so that the side with the rain-impervious, leather-like skin faced upward. The docks reeked of fish and brine.

Luke's ancestors were among thousands of fishers who fished from the schooners and dories plying the treacherous waters of the Grand Banks, just to the south of Fortune. The dorymen fished in bitter cold, often in thick fog. It was a dangerous occupation, indeed. When storms hit, everyone had to face the towering waves on his own.

Consider this log entry from Captain Walters when the famous *Bluenose* was hit by a storm on the Grand Banks on September 16, 1935: "The day begins blowing hard; bad sea running; sky looking very heavy. Wind hauled WSW with hurricane force . . . Vessel labouring very hard and terrific sea running . . . Vessel pounding very heave aft. Impossible to do anything . . . Tons of water going below doing the damages, causing vessel to leak very bad. Hard to keep continue pumping. So ends this say."

Many Grand Banks fishers were not as lucky as Captain Walters who saved his famous vessel, which limped into port one week later.

Luke was a little boy when the change came. In the 1950s and 1960s, industrial fishing started up on the Banks: all that steel, ice, and the weighted nets that went on forever. They marked the beginning of the end. In a few short decades the inshore fishery was wiped out as the cod were overfished to near extinction. Outport communities in every bay around the province were dealt a death knell. A provincial economy, built on a single resource, had been destroyed forever.

Once Luke made more than $100,000 a year fishing cod. Today he makes less than $30,000 fishing lobster and shrimp on the Grand Banks, scouring the shallows for fish the cod once ate, scouring and tearing at the habitat that once served as spawning grounds for the immature cod. "I'll be out of this within five years," he says. "I'm going into tourism. Fishing is no good anymore."

"But who's to blame for the cod collapse?" I ask.

"Blame enough to go all around," Luke says. "And who knows, maybe one day the cod will come back. But it will never be the same. And lots of people will have to change their thinking. There's an old saying on this island: 'If it runs, walks, or flies, kill it.'"

A brave and honest man. Many others are far more obtuse when asked who killed the Grand Banks cod, which is ultimately responsible for the near extinction of one of the world's most important renewable resources. To a visitor, many of the responses resemble an updated version of the "good German" defence: by just doing their job, paying the bills, going along to get along, people did not have to consider or take responsibility for the slaughter at sea. If everyone on the island is responsible—and the slaughter was sanctioned, torqued, trumpeted by politicians, officials, industry leaders, and the on-and off-shore fishermen themselves—then, it can be falsely argued, no one is responsible.

Meanwhile, many have fled the province. In 2005, Barry Shortt moved from Burnt Point, a village off Highway 70 that faces the North

Atlantic along the north arm of Conception Bay, to Leduc in northern Alberta, and now works as a mechanical engineering technician doing piping design for refineries and oil sands plants. Shortt bought a house, got engaged, and is expecting his first child, so it would be difficult to move back to the East Coast. "I'm not interested in going back to Newfoundland because you have to look at the opportunities and the position Alberta is in right now," he says.

Others who have been here for decades consider Alberta their home and no longer feel a strong connection to Newfoundland. Dominic House, now fifty-four, came to Fort McMurray from St. Alban's Bay in 1979. "This is home for me, it's where my kids were born and Alberta has been good to us," House says. His first job was as a truck driver. Eventually, he worked his way up to a job with the human resources department of the same company. He would never think of moving back to the East Coast, he says, for a job that would likely be temporary. "You're looking at a lot of jobs for the construction phase. If you have a permanent job in Alberta, why would you leave?"

Others, though, dream of returning to outports such as Fortune. A promise made by people with the sea in their blood. Today, working the oil sands of northern Alberta, they make their plans and pray—against all hope—that they will once again fish for King Cod on the Grand Banks.

8

Requiem for the Beothuk

"One story has it they were hunted like dogs…"
—Professor Michael Temelini

As a coda to the mystery of who killed the Grand Banks cod, a visitor confronts a very human puzzle.

Newfoundland was once peopled by a tribe known as the Beothuks. Somehow, every last man, woman, and child have been killed. The tribe has been exterminated, and some call it genocide.

A visitor, wending his way up the steep San Franciscan hills starting at Hill o' Chips Street—a short, steep, winding road that connects the east end of Water Street with the eastern end of this ancient city—passes the pubs on George Street before pausing in front of the now-empty site of Canada's most infamous orphanage, which closed its doors in 1989. For almost a century, the Mount Cashel orphanage was a symbol of Christian charity until it became synonymous with the terrible physical and sexual abuse inflicted on its residents by members of the Christian Brothers.

Our destination is The Rooms, home to the Art Gallery of Newfoundland and Labrador, the Provincial Archives, and the Provincial

Museum. Opened in 2005, The Rooms is named after—and designed to resemble—the traditional fishing "rooms" or buildings that dotted scores of outports throughout Newfoundland's fishing past. The facilities include a breathtaking view of both the city below and the Atlantic Ocean in the distance.

The Rooms have also drawn criticism to the Newfoundland government for its insistence that the building be erected over the former Fort Townshend, an 18th-century fort that is now buried under asphalt. Archeologists and historians said the complex would destroy whatever was left of the fort's foundation, which they would have preferred to see left for future excavation. In the end, the government prevailed.

The place now hums with activity. In one section, people throng to check the genealogy of their European ancestors—for free (this is one of the museum's most popular attractions)—those who conquered and settled Newfoundland: primarily the English, Irish, and French. In late summer, hearing the accents—Dublin by way of the East End of London—one could well imagine a past of schooners, dories, and salted cod.

It's a reverie shattered by a startling realization: Were there no First Nations here? If there were, what happened to them? And, why is their story so underreported? For a visitor curious about the fate of the now-vanished Beothuk tribe of Newfoundland, little information is to be found in The Rooms.

Memorial University political science professor Michael Temelini has noted the singular lack of information when it comes to Beothuk history, leading him to wonder whether this small First Nation had been the victim of genocide. "One story has it they were hunted like dogs—for fun," he said in an interview. "It may not be true, but I would like to know more."

Like all Newfoundland history, the story is complex and nuanced. There are few reliable descriptions of the Beothuk, and those that do exist conclude that by the mid-1500s, Newfoundland's native population tended to avoid contact with outsiders. Within 300 years, the tribe had disappeared altogether.

One particular book, *A History and Ethnography of the Beothuk* by historian Ingeborg Marshall, is today considered a standard reference on the Beothuk. During a speech given in St. John's in 1996, Marshall traced the relationship between the Beothuk and English settlers, from their initial peaceful contact throughout the years of hostility that, ultimately, led to the extinction of Newfoundland's First Nation some two centuries later.

Marshall began by describing letters by early English colonists John Guy and Henry Crout that told of a meeting with the Beothuk at Trinity Bay in 1612. Guy and Crout wrote of witnessing a Beothuk ritual involving shaking a wolf skin, singing and dancing, and striking their chests, which was followed by the Beothuk and the colonists exchanging gifts, sharing a meal, and trading peaceably through a silent barter system. Guy noted the Beothuk appeared harmless and that he planned to return to them the following year.

In Crout's letters to his employer, mining magnate Sir Percival Willoughby, the colonist described returning to the Beothuk camp the following year only to find it deserted. He continued to trade for furs elsewhere and while he occasionally glimpsed Beothuk in the woods, they would not approach him. Crout revealed he had to prohibit his men from attempting to catch the Beothuk by force, as he felt certain the natives would be "bent to revenge" should they be wronged in any way.

Marshall's speech went on to explain that relations between the Beothuk and the English steadily deteriorated over the years. In 1613, the natives gathered to meet with John Guy, but when the crew of a passing fishing boat shot at them, they immediately fled and went on to do "much mischief" in Trinity Bay. According to Marshall, the situation was exacerbated due to European encroachment and skirmishes with the rival Micmac tribe, resulting in the Beothuk gradually losing access to their traditional campsites.

In the 1720s, English settlers expanded their fishing operations into Notre Dame Bay and the Beothuk responded with violence, later shooting at English trappers who were quick to fire back. While these are the earliest recorded killings, Marshall reminded her audience that it's conceivable that Beothuk members had earlier died at English hands and these incidents simply hadn't been recorded.

During this decade, the Beothuk also battled with the Micmac, who eventually ousted the Beothuk from their coastal lands, substantially reducing their overall territory and curtailing access to their traditional foods.

According to Marshall, historical records make no mention of the Beothuk again until 1758 when a Beothuk woman and child were killed and a young boy was captured. This incident may have led to the Beothuks' subsequent killing of a shipmaster and five of his men in the Bay of Exploits, and also to the deaths of the first white settlers in Hall's Bay. So began a vicious cycle of murder and revenge where, says Marshall, the Beothuk usually emerged as the loser.

By 1768, Newfoundland governor Hugh Palliser had heard several tales of English brutality, spurring him to send Lt. John Cartwright up the Exploits River to make peace with the Beothuk. Cartwright never managed to locate any of the tribe, possibly because it was summer and they were on the coast, or because they had gone into hiding. He estimated the Beothuk population then to number about 300 to 500.

Violence continued, following Cartwright's unsuccessful expedition, with the Beothuk killing fishermen and the English retaliating with more blood. According to Marshall, perhaps the most disturbing tale recorded is of the "Peyton raid," in 1781, when an English trader, John Peyton Sr., and two men arrived at a Beothuk camp and began shooting indiscriminately. They then pursued the fleeing Beothuk, beating to death one man who was unable to escape.

Interestingly, Marshall noted that this violence is in sharp contrast to the positive experience of four French sailors six years later. Shipwrecked near Shoe Cove, the men were taken in by a group of Beothuk until a French ship eventually found them. According to the sailor who recorded the event, a teenaged Beothuk girl took a fancy to him and he was even considering staying with the tribe permanently before his rescue.

In 1784, having heard of the various atrocities against the Beothuk, prominent trader George Cartwright of Labrador began to lobby for the tribe's protection. He proposed the establishment of an Indian Reserve and predicted that the Beothuk would not survive unless the government took action. He even offered his services as an Indian agent, but his proposal fell on deaf ears.

Cartwright wasn't alone, as the like-minded Captain George Christopher Pulling also called for protection of the Beothuk. As evidence, he recorded a number of violent acts against them, which Chief Justice Reeves used to take up the cause in 1793, when he pleaded for a change in parliamentary policy towards the Beothuk.

The government still refused to budge, noted Marshall. By the turn of the 19th century, it was clear to British government agencies and Newfoundland governors that the situation had become untenable.

One step the governors took was to offer a cash reward to anyone who could bring a Beothuk to St. John's, in the hopes of employing the native as a kind of ambassador and developing friendlier relations and an understanding with the tribe. In 1803, furrier William Cull captured a Beothuk woman and brought her to St. John's, where she was given gifts for her people before the governor ordered Cull to return her to her home. In Marshall's account, however, Cull simply left the woman in the forest, as he feared entering the Beothuk camp, and it's unknown whether she made it back to her people.

Relations between the Beothuk and the English remained tense and, in 1811, events took another dark turn. Commissioned by Governor John Thomas Duckworth, Captain David Buchan and a party of marines set out on a peaceful mission to find the Beothuk—although they went well armed. The group came upon a settlement at Red Indian Lake in the dead centre of Newfoundland early one morning. At first, the Beothuk accepted the visitors without incident and, feeling reassured, Buchan returned to his camp to gather gifts for the tribe, leaving two of his men behind. However, the Beothuk grew suspicious in his absence and they attacked and killed the two marines before fleeing into the forest. Buchan searched for the Beothuk during the next two summers but couldn't locate them. By then, said Marshall, only seventy-two people remained in the tribe.

Another deadly incident followed, in 1819, when local merchant John Peyton Jr. requested permission from the next Newfoundland governor, Sir Charles Hamilton, to retrieve his stolen property from the Beothuk, who had been pilfering his fishing boats. Governor Hamilton not only granted permission but also encouraged Peyton Jr. to bring back a Beothuk who could help to establish friendly relations between the groups. Events didn't turn out as peacefully as Governor Hamilton had hoped, reported Marshall. At a Beothuk settlement, Peyton Jr. captured

Chief Nonosabasut's young wife, Demasduit, who was weakened from having just delivered a baby that died shortly after. When Nonosabasut tried to rescue his wife from Peyton's men, he and his brother were shot to death. Peyton Jr. was brought before a grand jury for the killing but was later acquitted of any crime.

In a *Toronto Star* article published on August 6, 2005, reporter Peter Calamai wrote that the killing of the chief and his brother, the Beothuk's two remaining leaders, sealed the fate of an entire tribe and unique way of life. While accounts differ as to what actually happened that day, Calamai reported that a History Television documentary being made on the Beothuk using new forensic evidence had brought to light a possible reason why the encounter had turned so deadly. In his article, Calamai quoted forensic anthropologist Tracy Rogers, who examined a reconstruction of the chief's head, a replica skull, and a 3-D scan. She found no evidence of trauma that could have led to Nonosabasut's death, but she did discover damage to the lower jaw that had healed over time, perhaps caused by a blade. According to the documentary producer, Christopher Gagosz, the blade was from a bayonet wound, possibly a wound delivered by John Peyton Sr. of the infamous "Peyton raid" of 1781. "Accounts suggest that several years earlier, Nonosabasut may have had a violent run-in with another group of armed settlers led by Peyton [Sr.]," reported Calamai. The elderly Peyton was also on the 1819 expedition with his son, John Peyton Jr., and it's conceivable that it was Nonosabasut's second encounter with Peyton Sr. that led to the fatal shooting.

Whatever truly happened that day, the captive and newly widowed Demasduit's arrival in St. John's caused quite a stir. In her speech, Marshall read out a report by an anonymous writer published on May 27, 1819, in the *Mercantile Journal*:

> On Sunday last, the curiosity of the good people of this town was gratified by an unexpected visit from one of the Red Indians, a young woman, about twenty years old. In consequence of the habitual persecution and cruelty, which every well-informed person in this island knows to have occurred, we could not but believe that the Red Indians were the most ferocious and intractable of the savage tribes. And it is with no less astonishment than pleasure that we find in the young woman, which has been

brought amongst us a gentle being, sensibly alive to every mild impression and delicate propriety of her sex. Is it not horrible to reflect that at the very moment, while we set down at our fire sides in peace and composure, many of her country men, in all probability as amiable and interesting as this young woman, are exposed to wanton cruelty . . . We might remember that as far as priority possession can convey a right of property, the Red Indians have the better title to the Island.

According to Marshall, this was the first public admission of the Beothuk's legitimate claims to the island of Newfoundland. Some of St. John's citizens formed a committee and financed a mission to return Demasduit to her people, but she died in 1820 of tuberculosis before she could be brought home. Captain Buchan carried her remains to the settlement where she had been captured and tried, unsuccessfully, to find any of her tribe.

In 1823, three starving Beothuk women were discovered inland in a wooded area by furriers, a young woman named Shanawdithit, along with her mother and sister, who both died shortly afterwards. Ironically, Shanawdithit was taken in by the young John Peyton Jr., now a magistrate, to work as a servant in his household on Exploits Island, where, said Marshall, she "seems to have been reasonably happy."

A year earlier, a scientifically minded entrepreneur, William Eppes Cormack, had trekked across the entire island in an unsuccessful search for members of the Beothuk tribe. Undiscouraged, Cormack founded the Boeothick Institution in an attempt to solicit public support for his plans to rescue the remaining tribe.

Cormack had Shanawdithit brought to St. John's in 1828 in order to learn as much as he could about her people. Shanawdithit told him of how their numbers had dropped from seventy-two in 1811 to only a dozen at the time she was found by the furriers. She was pessimistic about their chances of survival since there were not enough of them to maintain the caribou fences and having been driven away from their coastal lands meant they had been cut off, in her own words, from "their means of existence."

Although Cormack had hoped Shanawdithit could help him to establish peaceful relations with her people, she resisted. According to Marshall, she told Cormack that "from infancy all her people were taught to cherish animosity and revenge against all other people; it was enforced by narrating the innumerable wrongs inflicted on the Beothuks . . . and that if the Beothuks made peace and talked with them, they would not, after they died, go to the happy island and hunt in the country of the good spirit."

Shanawdithit, the last known remaining Beothuk, died of tuberculosis in June 1829 in a St. John's hospital.

Marshall recounted that it didn't take long for public opinion on the Beothuk to change following Shanawdithit's death. In an 1832 *Royal Gazette* article quoted by Marshall, the Beothuk were described as having had their land and resources taken from them unlawfully, while the circumstances of their demise were blasted as "repulsive." The Beothuks had gone from being considered harmless trading partners in the 1600s, summarized Marshall, to "dangerous and sub-human" in the 1700s, before being acknowledged as a people needing protection in the late 19th century. And finally, they had become "victims of prejudice and cruelty" in the public eye once all had vanished.

Marshall concluded her speech with the following statement: "While I agree that both prejudice and cruelty were at work, I suggest that the Beothuk had not been powerless victims who had allowed circumstances to rule their lives. In my view, they were a heroic people who valued their independence and traditions above all and were prepared to face hostilities rather than be subjugated."

Today, many First Nations vehemently disagree with Marshall's contention. They claim that the Beothuk were indeed victims of genocide.

In 1997, as Canada celebrated the 500th anniversary of the "discovery" of Newfoundland and Labrador, aboriginal protesters came out to demand justice and to commemorate the genocide of indigenous peoples.

First Nations demonstrators trailed Queen Elizabeth II throughout her tour of Canada. The Queen had travelled to Canada to be part of Cabot 500, a multimillion-dollar celebration of the Italian mercenary sailor John Cabot's voyage to Newfoundland. Cabot arrived there in his ship, *The Matthew*, on June 24, 1497, signalling the beginning of the British and French assault on the land and indigenous peoples of North America.

Born Giovanni Caboto, Cabot was sponsored by the English King Henry VII. Henry empowered Cabot and other mercenary sailors to "search out territories of heathens and infidels and to enter and seize them . . . to occupy, possess and subdue [them] . . . as our vassals." Explorer John Cabot's landing, the Queen said in a speech, represented the geographical and intellectual beginning of modern North America.

What the Queen did not say was that less than 350 years after what First Nations describe as "her pirates and thieves" invaded the sovereign Beothuk territories in Newfoundland, the Beothuk nation had been exterminated.

The other indigenous populations in Newfoundland and Labrador have come perilously close to extinction since the Europeans' arrival. Innu leader Katie Riche told a CBC radio reporter that day: "I see nothing to celebrate. Along the way a whole nation, the Beothuks, were wiped out. We don't want that to happen to us."

This sentiment is echoed in the 2001 book *The Beothuk Saga* by Bernard Assiniwi and Wayne Grady, which traces the rise and fall of the tribe. The book's somewhat complicated narrative powerfully tells the story of how the Beothuk were quickly betrayed by the English, who used their political trading savvy to fracture the tribe's hold on its territory. The final section of the novel, entitled "Genocide," chronicles the butchering of the Beothuk by their callous conquerors and explores the plight of the dying tribe.

What can a visitor make of these anthropological and historical contradictions? What happened to this tribe? And, why does it seem to be a public embarrassment?

Walking back down the hill from The Rooms, one is left with a new and unsettling series of questions about the people who settled this rocky shore with its barren heaths and rolling hills of stunted pine. Settlers from the British Isles and other parts of Europe came here to fish for cod. Brave beyond any definition of the word, they fished the Grand Banks in schooners and dories. To survive in this place, fierce men learned to stalk and kill cod, moose, and seals. And, for any dereliction of duty, British military justice was instant, tough, and cruel.

Still, a visitor is left to wonder: Did such a mindset lead to the genocide of the Beothuks who were outsiders seen as a nuisance to a fishing society? The thought will not go away. Can it be legitimate to draw a parallel between the extermination of a First Nation and the slaughter of the Grand Banks cod? It happened so long ago, during a time of brute imperial force, when aboriginal peoples around the world were dismissed as heathens, unpaid labour, and less than human. Three thousand miles to the west, in January 1887, Nisga'a chiefs arrived in Victoria's inner harbour. Determined to settle the Land Question, they were met by Premier William Smithe, who barred them from the Legislature with this rebuke: "When the white man first came upon you, you were little better than wild beasts of the field."

Still, the parallels seem to me to be very real and shameful—and instructive for our time. Why has this aboriginal heritage been buried in the official story? Buried, like the archeological sites on which the museum was built; covered up and sealed under the asphalt of official history.

9

Pacific Salmon: Going the Way of the Grand Banks Cod?

Back in another time, young people began to ignore the warnings of their elders. They would kill animals needlessly and began to do the same with the fish they caught, maiming them and throwing them back in the lake. Now, in those days, animals, fish and birds were known to have supernatural powers and so the elders lived in constant dread of the catastrophe they knew would happen.

One day it began to rain. Soon all the waters rose and covered the world. Many people perished. Among those still alive there was an old woman who had an only daughter.

"This is the great catastrophe that we have been expecting," said the old woman to her daughter. "It has been caused by the thoughtlessness of our young people, and as you were always mindful of what I said, I am going to save you while I can. The others shall be destroyed because the waters are continually rising." As legend tells it, the old woman and her daughter were among a handful of people who survived the flood.
—Nisga'a legend

When the Grand Banks cod collapsed in 1992, a warning shot ricocheted around the world. Four thousand miles to the west, out on Canada's Pacific wild salmon fishing grounds, experts such as University of British Columbia professor Carl Walters warned there were far too many troubling similarities between salmon and the cod. And furthermore, unless a comprehensive corrective was set in place immediately, the iconic Pacific salmon was doomed to eventual extinction. Today, in 2008, Walters' dark and apocalyptic prediction is coming to pass.

Like the Atlantic cod before it, is a complete collapse of the Pacific salmon fishery industry now in the cards? British Columbians, at least those who make their living off salmon, don't sound optimistic. Just ask Mike Forrest.

Growing up on the south bank of the Pitt River near its convergence with the world's greatest salmon resource gave Forrest a unique sense of kinship with the ebb and flow of life along the Fraser River. Forrest's grandfather settled here and began commercial gillnet fishing for sockeye, chinook, and other Pacific salmon species in the 1920s. Mike's father took up the same vocation, and by the time Mike was a teenager, he could pilot a boat, set and retrieve a net, and make a living selling his catch to local canneries.

Anyone who wanted to fish could buy a commercial licence and a boat, and barring a few restrictions, go out and set a net during any month of the year.

It was, as he understood it, his heritage to be a Fraser River salmon fisherman. In 2008, at fifty-nine years of age, he cannot let go of the anger, bitterness, and frustration he feels about the collapse of a fishery that began in earnest with the arrival of settlers who followed the 1858 Fraser River gold rush.

The first cannery opened here in 1871, predicated on the presumably limitless numbers of fish moving each year into the river. By 1903, the industry had played such a fundamental role in the development of the British Columbia economy that the city of Vancouver incorporated a fisherman into its first coat of arms and underneath the illustration added a motto: By Sea and Land We Prosper.

Of course, members of the Sto:lo tribal group and other Coast Salish First Nations up and down the lower and interior reaches of the Fraser had been fishing the mainstem and its tributaries for millennia. The abundance of fish was so great that it delivered to a people limited to Stone Age technology the luxury of time to develop one of the world's richest aboriginal cultures.

Their labour-intensive methods, including dip nets, weirs, and trawl nets dragged between two canoes, barely made a dent each year in the spawning migration of millions and millions of salmon that began arriving in spring and continued on into the winter months. Just as in the waters off the coast of Newfoundland and Labrador hundreds of years ago, the Fraser once teemed with fish, and the supply seemed endless.

The salmon provided a rich source of protein for aboriginals in the Lower Fraser close to Vancouver, for those living on muscular Fraser tributaries such as the Thompson River, and on the Fraser mainstem itself—and even in remote villages farther north as far as the British Columbia interior forestry hub of Prince George, which is almost 800 kilometres from the point where the river empties into the Strait of Georgia.

At places all along the river, prime fishing spots have been passed along from generation to generation of aboriginal families.

For people like Mike Forrest, the at-large commercial fishery represents a similar inheritance. He recalls one incident from far back in his own childhood when he and his grandfather, Captain Sam Forrest, were confronted by a fisheries officer for fishing during a dense fog that posed a navigational hazard on the river. "I remember getting caught up in the fog and a fishery officer coming alongside and saying, 'Sam, you're not supposed to be out here. It's illegal.'

Sam said to him, 'Y'know, my family's got to eat and I have to catch some fish for them.'

"That's where I come from and that's my heritage in this fishery."

These days this heritage is more a function of memory than an exercise of Mike's skills. Between 1999 and 2007, Forrest and about 380 other licensed Fraser commercial fishermen were obliged to keep their boats tied up at the docks and forgo the sockeye during their annual migration into the Fraser.

The most recent year, 2007, was the worst on record. With October nearly past, Forrest was still awaiting the possibility that the Department of Fisheries and Oceans might consent to open the river for a few days to allow the commercial sector to fish for chum, one of the lowest-grade species of salmon.

It was evident almost from the start of the 2007 migration that Fraser River fishermen would not get an opportunity to fish sockeye, the most prized and highest-priced commercial species due to an exceptionally low return. Things are not expected to improve in 2008, the weakest of the four recurring cycle-years that form the Fraser sockeye's biological calendar.

Nor are things expected to improve much in the future, as the fish continue to show signs of exceptional weakness under the weight of an increasing number of problems associated with climate change—including warmer water, which is depleting the Pacific Ocean food chain and curtailing survival of all species of salmon as they mature to adulthood.

Warming temperatures are also a threat in freshwater. In recent years, the summer temperature of the Fraser River has been a lethal danger to sockeye as they attempt long-distance migration to natal rivers and creeks.

Several times since the mid-1990s, summer sockeye have baffled scientists and fishermen alike by arriving at the river mouth several weeks ahead of their normal migration time. Instead of proceeding upstream, they have milled there in the hundreds of thousands, eventually falling prey to disease or commencing migration in a severely weakened state, and prompting severe conservation concerns about the viability of the species.

Convergent with this collapse has been a profound and fundamental reconfiguration of the way in which the fisheries department regulates the commercial fleet and the aboriginals who take fish from the Fraser watershed.

In 1990, a groundbreaking Supreme Court of Canada decision favoured the right of a Musqueam First Nation fisherman, Ron Sparrow, to fish for food, societal, and ceremonial purposes. That decision stripped away from the fisheries department the authority to assign just a nominal allocation of fish to aboriginals in the Fraser watershed—and to aboriginals in other fisheries and in other culturally based activities here and across Canada.

Aboriginals could not fish as an explicitly commercial enterprise in this new fishery, but were allowed to catch a greater number of salmon in the exercise of their constitutional rights. As a consequence, the non-aboriginal commercial catch was reduced in order to maintain the department's conservation objectives for spawning salmon.

Bands such as the Musqueam and Tsawwassen have some special commercial fishing arrangements outside of their constitutionally protected ones—but those arrangements are contingent upon the opening of a wider commercial fishery that would support people like Mike Forrest.

Except that there aren't enough fish to let either group on the water to fish for profit.

What really rankles Forrest, however, is the inconsistent application of rules the department uses to manage the fishery and ostensibly fulfill its primary mandate to protect the fish. Fifteen years ago, Mike Forrest and his brother Raymond purchased and upgraded a $350,000 aluminum boat on the assurance of the federal government that a licence buy-back scheme would reduce the competition on the river to a point that would provide a decent income for those who remained. Those promises were empty, he says.

"We invested in a big boat based on promises by various people. We were told there was a future in the fishery for those people that stayed in, that they were part of a licence limitation that would have benefits in the future. And then, of course, the benefits were all given to the aboriginal people and that boat has fished 100 days in fifteen years."

Commercial fishermen and aboriginals seeking to fish for profit were kept off the water in 2007 on the premise of a conservation concern for sockeye—except that the aboriginal fishery for food, societal, and ceremonial purposes, otherwise known as the FSC fishery, was allowed to continue for bands in the lower Fraser.

For example, the population of Early Stuart sockeye, an early arriving bellwether strain of fish destined for the far upper reaches of the Fraser, numbered 13,000, and the department allowed First Nations to keep fishing despite an evident conservation threat to that salmon group.

By contrast, when it became apparent in the mid-1990s that the spawning population of Thompson River coho had plunged to 30,000, the department ordered a broad-based sport-fishing closure that took 100,000 sport fishers off the water for good and cost the British Columbia tourism economy more than $200 million in lost revenue every year since then.

"This is more involved than average people are willing to listen to," Forrest says.

"The natives were allowed a bycatch, if you will, of Early Stuart sockeye with only 13,000 through Mission when we were looking for 50,000. It was a real bone of contention with us that people fished into conservation and are allowed to by DFO."

Forrest, a city councillor in his hometown of Port Coquitlam, is unabashed in his assertions that decisions about fishing in the Fraser are based on race first, and conservation second, a stance that has made him one of the most controversial figures in the whole debate about the future of the British Columbia fishery.

"In the sixties, we had various returns at the same kind of level in total sockeye return we have now. We had various fisheries—and the runs did fine. We had spawners, we had replenishment, we had great runs into the 1980s—at a time when we had a heckuva lot more fishermen. We had no licence limitation at those times."

"Now we have only 370 to 380 licences or something [like that], and we have all these restrictions. We have stories about trouble with various stocks. For us the thing that has changed is the aboriginal fishery. Nobody at DFO likes to hear that. The general population thinks we're a bunch of racists for saying so. But that's the difference in the last fifteen years."

One thing that has changed, and is not reflected in the commercial sector's denunciation of DFO's management strategies, is that there is far less certainty about the number of spawners that is needed to be put on the fishing grounds in order to guarantee a decent return of fish four years down the road when the offspring of those spawners return.

In the 1960s, and even as recently as 1990 when twenty million salmon returned to the Adams River, it was possible for the commercial fleet to intercept as much as 90 per cent of the spawning run. DFO and the fishermen could rely on a rich ocean food supply to sustain a healthy recovery for the offspring of those fish that escaped the nets and spawned.

Crashing ocean survival levels mean those expectations can no longer be met, making the risk of overfishing a run increasingly perilous to the viability of the sockeye species in general, and the populations of many individual Fraser tributaries in particular.

Forrest believes that a commitment to conservation, as opposed to lip service and inconsistent rules, could turn the situation around before it reaches a crisis of Grand Banks proportion, but he's not counting on DFO to make any significant adjustments in its management. He wants an opportunity to fish, but he also wants the fish to thrive for future generations.

We're having meetings right now with DFO trying to get somebody to listen to reason about the balance of conservation. If you have a conservation problem, it should be that everybody has the same conservation problem instead of two different sets of rules; otherwise, we will not get the stock to survive.

This aboriginal fishing strategy process has two sets of rules. The department keeps saying it has one set of rules. Sorry, it's very obvious to us on the water that it's different than that.

It is DFO policy to not affect the native community and a whole bunch of people. They don't want to affect treaty issues. The only way to stop a fishery is to prove in court there's a conservation concern.

DFO will state a conservation concern to stop us from fishing because they don't have to go to court about it. But they will not state a conservation concern to the aboriginal folks. It won't stick unless they have the numbers to prove that this stock will not survive or will not replenish if the natives go fishing.

We need to rebuild the Fraser stock. Let's agree on an escapement number for the amount of fish we need to put on the gravel, and have everybody work to that number.

"It's not just a problem for fishermen," Forrest adds.

Streams in urban areas are increasingly challenged by development, pollution, and the volatile new cycles of flood and drought brought about by climate change.

Beyond the Fraser Valley and above the Fraser Canyon, both spawning and hatchling salmon are in competition for water with cattle ranchers, hay farmers, and other licence holders, who in many areas have enough authorizations to remove more water than is actually available from a given stream.

Federal and provincial fisheries agencies have both spoken about the need to conserve water for streams, but as Watershed Watch Salmon Society noted in a recent report, nobody has a solution to the problem because water use is so dispersed across the Fraser basin.

"It's so easy to stop us," says Forrest. "The natives are very difficult to stop and therefore it's not done. The water and habitat issues for various creeks are less obvious to the average population but are very concerning and everybody needs to be part of it—the sport fishers, the ranchers, the general population. We've got some buy-in with that, but we haven't got lots."

STO:LO tribal council fisheries adviser Ernie Crey shares Forrest's disappointment with DFO's management of the fishery, but little else.

At fifty-eight, Crey is a former employee of the fisheries department whose soft-spoken and articulate analysis of aboriginal fishing rights has rendered him a primary figure in the public debate on the issue.

By his account, the Sparrow decision in Supreme Court was such a fundamental shift in favour of aboriginal rights that it left the fisheries department with few options. The courts, he notes, "were silent" on the issue of how many fish the Indians should be allowed to keep.

Crey says that in order to maintain some semblance of management over the fishery and serve its primary mandate to conserve and protect salmon, the department created a two-tiered system for aboriginal fishing activity.

Tier one is the food, societal, and ceremonial fishery, which is a constitutionally enshrined upgrade of the old Indian food fishery.

Tier two is the aboriginal fishing strategy, the AFS, which binds commercial fishing activity as an exercise of aboriginal rights to the fishing opportunities of the commercial fleet.

"We don't get to catch and sell salmon when there are no opportunities for members of the industrial fishery," says Crey. Moreover, the Fraser salmon fishery "was already in decline" before DFO ever struck the first new-era relationship with aboriginals, Crey adds. "There have been various licence rationalization programs amongst the commercial fleet for decades prior to that. Even with the announcement of the AFS, part of the overall strategy was to buy back commercial licences."

Crey says you don't need a PhD to recognize that the problems now facing the fishery are more fundamental than the debate over aboriginal rights. Even many First Nations are facing the loss of their constitutionally enshrined food, societal, and ceremonial fisheries amid the ongoing collapse of sockeye. He notes, "I recognized several decades ago that we were coming to the crossroad and that decisions would have to be made, and so did government. In 1982, Peter Pearse headed up the Royal Commission of Inquiry into the Pacific coast fishery. He had a raft of recommendations for the government of Canada that would lead them down the road to better-managed fisheries."

Among those recommendations, made a full ten years before the Grand Banks cod disaster, was a common-sense observation that there were too many boats chasing too few fish and that the future of the salmon fishery was contingent on a reduction in the number of commercial fishers who had access to it.

"People were realizing then that the fishery was undergoing a dramatic change and that there was more on the horizon," Crey says.

But even as far back as 1970, when he was a young social worker hired to work in a small aboriginal village on the west coast of Vancouver Island, it was evident to Crey that there was trouble ahead: "I wasn't a professional biologist or someone who understood the commercial fishery well. But even then I could take one look around the village and I could see that the commercial fishery was on the decline. The dock for members of the band where they would normally dock their commercial vessels held only one commercial vessel.

"There were several others they owned that had been scuttled in that bay in front of the Indian village, directly across from the village of Ucluelet, the non-aboriginal community where there was a fish plant."

"There were dead boats and people on welfare, waiting for the fishery to come back. It never did."

Crey collected further insights in the 1980s when he went to work for the fisheries department. The department's economists were already anticipating the economic cost to Canadian taxpayers of ignoring Pearse's recommendations about reducing the size of the British Columbia salmon fishing fleet.

When I was at the department in 1984, '85, '86, I was in more than one meeting in which fisheries planners and managers sat down and discussed how costly it is to manage the fishery on the Pacific coast, particularly the salmon fishery.

They looked at the cost to run their operation as the department in the Pacific region versus the economic return on the industrial fishery. They forecast that at one point in the future the cost to manage the fishery would exceed the economic return to the commercial fishers and the processors. They showed where the lines would cross.

"We're there."

Even then, before Ron Sparrow and the Musqueam initiated their constitutional challenge to DFO's methods of regulating the old Indian food fishery, the department perceived a new and stronger role for First Nations—even if nothing happened except the passage of time.

That was another exercise we went through in the department. We looked at population growth of aboriginal communities in the Fraser watershed and forecast what the population might be given the rates of growth in the eighties. Then they estimated what the drawdown would be on the sockeye run by those communities, off into the future. We worked the calculation to the point at which the aboriginal population would be at such size that there would be no sockeye for any other interests.

"Guess what? We're there."

Crey's recollections are borne out by Statistics Canada, which reported that the B.C. aboriginal population is growing at a rate twice as fast as the population at large, and the trend is expected to continue for at least another decade beyond 2007.

Crey is quick to caution, however, that the aboriginal entitlement to fish is in peril in many places—notably in the upper Fraser watershed beyond the main B.C. population centre—due to declining fish populations.

"My take of what happened last summer is that the sockeye runs to the upper reaches of the Fraser River watershed—although there may be some good returns from time to time in the future—are all but finished. What this means is that aboriginal fisheries such as Hell's Gate or Lytton are going to be relegated to history because there will be very few opportunities for those people to fish in the future."

Such a fate comes as no surprise to Crey after nearly four decades of observing B.C. salmon. "As the years have gone by I can see the fish are not showing up in the numbers they once did and, in fact, there were some seasons I went out fishing that I could see the fish meandering around and swimming in circles, and you could catch sockeye in the Fraser canyon just by reaching in and taking them in your hand. You didn't need a net.

We are just in for more of it. The signs have been there and what were once signs are simply now the loud alarm bells like one would hear walking down a hallway in a school where the fire alarm was going off.

It hurts your ears, you can't ignore it, and you've got to get out.

"I see a lot of hardship in the future for half of the province's First Nations folks. There are ninety-four registered bands in the Fraser River watershed. They may have places to fish, but there will be no fish to be had."

Crey qualifies his remarks by adding that he's "not a negative guy." But he suggests that over the past fifteen years there has been too much of a preoccupation among commercial fishermen with turning back the clock on the reconfiguration of the aboriginal fishery at the expense of an opportunity to consider ways to protect salmon runs in the face of climate change—and a profound failure in the fisheries department to do anything other than referee the debates among the various user groups along the Fraser.

"I've always taken a sort of upbeat attitude about sockeye returns to the Fraser River and how if everyone were nice to one another, worked together, quarreled in a reasonable fashion, that we could restore these runs, that there could be fish for everyone," says Crey.

We could have spent the last fifteen years working together and attempting to persuade the government of Canada and the government of British Columbia to devote more resources to restoring Fraser River sockeye runs, steelhead runs, coho runs, chum.

We could have spent the past fifteen years doing that. But, no, a lot of people just wanted to fight and argue. The ones that were smart in the commercial fishery diversified their licence holdings so they not only have licences for salmon, they have licences for the highly lucrative ground fishery and they are happy, making lots of money, doing their thing.

"Some feel they should just fish the lower Fraser and if they're not fishing, it's because DFO is a poor manager and all the fish are going to the Indians."

He adds that DFO cannot claim ignorance about the decline. There were a lot of studies with recommended courses of action—a lot of reports with recommendations that go back over thirty years and the government of Canada really hasn't acted on any of them. There is no vision there.

The department is subject to the shifts of political winds. If a powerful group shows up that shouldn't be out fishing but has a lot of clout at the cabinet table, it doesn't matter what a Paul Sprout [DFO Pacific regional director] or a senior fisheries management biologist might think is the correct thing to do based on science; he's overruled by the politics around the cabinet table because powerful interests want a fishery.

That's how they've managed the fishery and that's it. There are lots of good folks at DFO but that's how it works. The fisheries are driven by politics, not by science.

"That's the bottom line."

Any lingering doubt about the fisheries department's ultimate agenda was removed in the summer of 2006 by enterprising members of a Fraser Valley sport-fishing club and fisheries students from the British Columbia Institute of Technology, who stumbled upon a disastrous, DFO-supported gravel-mining operation on the bed of the Fraser River at a time when billions of pink and chum salmon fry were newly emerging from their spawning beds.

The BCIT students reported a great stink coming out of the gravel where the infant fish had died and begun to rot by the millions as a direct result of the gravel operation. An even greater stink subsequently emerged in Ottawa after it was revealed that fisheries officers, disgusted by the involvement of their own superiors in the death of the fish, had been witnessed standing on the bank of the Fraser, a copy of the Canadian Criminal Code in hand, discussing which sections of the Code might best be applied to a prosecution of their own agency.

The initial story broke nationally when one of Canada's most prominent environmentalists, World Rivers Day founder and Order of Canada laureate Mark Angelo, released results of an investigation that identified the fisheries department as a principal culprit in the gravel-mining operation that destroyed more than two million baby pink salmon nestled in the bed of the Fraser River in March 2006.

BCIT students and staff, along with members of the Fraser Valley Salmon Society, reported that the fisheries department failed to properly monitor a commercial operation building a temporary gravel causeway required to access an island being mined during the Fraser River's annual low-flow season.

Gravel mining is a controversial subject on the Fraser. Gravel flows down the Fraser year-round and begins to settle and collect once the river emerges from its violent canyon at the town of Hope, about 120 kilometres upstream of Vancouver.

Vast accumulations of gravel are exposed in late winter before the spring freshet sends greater volumes of water coursing down. Some communities in the northern part of the valley argue that removal of gravel deepens the Fraser's natural channel and reduces the risk of flood.

As well, gravel is enormously valuable to the local construction industry as a cheap, quick source of aggregate for concrete and fill. Some say there's no harm to fish if the gravel is mined after all the young salmon from that year's hatch have left. Others contend that the removal is as hypocritical as the so-called scientific research motives of the Japanese whaling fleet and that harm to young salmon, and to salmon habitat, is inescapable.

Both the BCIT group and the salmon society took photos documenting the manner in which the causeway rerouted the flow of the river away from a spawning area pockmarked with more than 6,500 pink salmon redds, or nests, over an area of several hectares.

Students digging in the gravel found countless dead alevin, which are salmon newly hatched and exceptionally vulnerable to environmental threats and predation, nestled there. Robbed of water, they had died.

Angelo, head of the BCIT program and the inspiration to several generations of British Columbia fisheries technicians and biologists, estimated that under normal circumstances, as many as 40,000 adult pink salmon would have survived from such a large field of eggs.

BCIT instructor Marvin Rosenau, a former provincial fisheries biologist well-versed in the brutal politics of gravel mining on the Fraser, told the *Vancouver Sun* on April 17, 2006, that the operation "blatantly disregarded" stream protection laws and suggested DFO staff would have been fully aware of the consequences of building the causeway. Rosenau, whose own battles with the B.C. environment ministry over gravel mining saw him removed from front-line duty in the fisheries branch, said he had no doubt that rank-and-file DFO staff had been instructed by senior officials in the department to ignore the situation.

"They would have had to be blind in order not to know the risk was more than trivial," Rosenau told the *Sun*'s Scott Simpson. Rosenau's remarks were even more pointed when he testified on June 1, 2006, before the federal government's standing committee on fisheries and oceans.

"The largest salmon run in British Columbia spawn in the gravel reach. Sometimes there are in excess of ten million fish. At least five listed species at risk are contained in that gravel reach," he said.

Rosenau testified that DFO's area habitat chief, Dale Paterson, "pooh-poohed" his suggestions that the causeway and the mining operation presented a threat to fish, claiming that they had in all likelihood hatched and migrated out of the area. Rosenau presented to the committee DFO's own data showing that "the pink salmon juvenile out-migration had barely started," and wasn't shy about pinpointing blame: "What I would say is that the Department of Fisheries and Oceans has turned around, and instead of protecting the environment, the authorizations have become political; they have become politicized. Executives and the senior managers are making decisions, and the local biologists and engineers are basically being cut out of the decision-making process. What you have is extreme habitat damage as a result."

Rosenau noted that at times during the mining operation there were various DFO employees on site, but they shrugged off suggestions about the threat to fish, saying the decisions were being made by senior bureaucrats. He said, "The thing that really brought it home was when one of the conservation and protection field officers said that DFO should be charged. They actually brought out the sections of the Criminal Code that DFO should be charged under. To us, it was absolutely stunning that one of the staff from DFO should be articulating that DFO itself should be charged, and not under the Fisheries Act, but under the Criminal Code for failing to meet their statutory responsibilities."

Frank Kwak of the Fraser Valley Salmon Society also testified before the committee and highlighted the apparent disinterest of the department in collecting fines it obtains in prosecutions under the Fisheries Act— Canada's oldest and most time-tested piece of environmental protection legislation. "The simple fact is that under the current legal system, there is no real requirement for anyone to pay their imposed fine if they do not choose to do so," Kwak told the committee.

"We are informed that if the outstanding penalty is more than $100,000, Justice Canada will look into taking action to collect it, but for amounts less than this, they do not. Just this last February, one of my organizations, the SDF [Sportfishing Defence Alliance], was informed, as a result of a specific request, that there is currently in excess of $1 million in outstanding fines for offences committed in fisheries in the Pacific region.

"This amount is up from $500,000 reported in 2003, and has more than doubled in two years. The fines go back to 1994 and range from $100 to $20,000."

Fisheries department Pacific regional director Paul Sprout, at a subsequent meeting of the parliamentary committee, said he believed the department made a concerted effort to arrive at a consensus on a gravel framework that involved a wide array of people. "We believe we took specific measures in 2006 to try to ameliorate the impacts of that particular site. But having said that, we discovered that there were issues in 2006, and we're going to learn from them and take them into consideration in the future."

In an interview in October 2007, Rosenau said the issue has been at least partly resolved by the revival of a committee that provides a more scientific perspective on Fraser gravel-mining operations: "However, this technical committee is now under extreme pressure to still provide gravel in locations that have no flood-control benefits, including removal from locations that are known sturgeon-spawning habitats, for example," Rosenau said.

Further, because this initiative is largely a political initiative to provide access of gravel to certain local interests, independent of any flood or erosion benefits, the issue is likely to be muted for only a short time until it rears its ugly head in a big way again . . .

"I would point out that in respect to Fraser River gravel removal clearly some of the conduct of DFO in the past has been flat-out illegal or, at the least, professionally negligent. However, where the issue will go now, no one is sure. Certainly there seems to be two groups of attitude within the public; one wants to get as much gravel out as possible, regardless of the environmental impacts and regardless as to whether the benefits are significant, because any benefit is viewed as a good benefit; the other group simply believes that this is a political scam to provide the development industry and friends of the local politicians in the eastern Fraser Valley with cheap and easy gravel and is using the fear of flooding to provide access to this aggregate."

Rosenau isn't the only expert with a disturbing story to tell.

Biologist John Werring, a veteran B.C. environmentalist and field investigator, recently prepared a report for the David Suzuki Foundation showing a widespread disinclination by the fisheries department to confront habitat damage, even after repeated entreaties by alarmed citizens.

"DFO is mandated under the federal Fisheries Act to do so. Yet our intensive field investigations have shown that the destruction of critical salmon habitat continues," says the introduction to the report, titled *High and Dry: An Investigation of Salmon Habitat Destruction in British Columbia.*

The Fraser gravel fiasco was part of the report, but Werring also reported on the alleged illegal and unchecked expansion of a salmon farm on the B.C. central coast; the destruction of a Class A fish habitat in Surrey, British Columbia's second-largest city; a logging operation immediately adjacent to a fish-bearing stream in the Queen Charlotte Islands off B.C.'s north coast; and several others.

A dmitting it was unable to protect wild salmon stocks and produce enough for growing numbers of commercial, aboriginal, and sport fishers, the Department of Fisheries and Oceans embarked on a techno-fix: It would artificially grow millions of baby salmon in massive cement hatcheries to solve the problem. Over time, the hatcheries have proved an expensive failure that leads directly to salmon death.

Once touted as the techno-fix for the declining number of Pacific salmon, some of B.C.'s hatcheries are now seen as a monument to government meddling with nature. In fact, according to new research, after thirty years of raising salmon artificially, well in excess of a billion dollars, government hatcheries may actually be responsible for a plummet in the population of prized wild coho salmon.

Concern for the commercial Pacific salmon fishery—an industry whose landed value has dropped more than 70 per cent to $60 million today—and helps to support hundreds of seasonal fishing and processing jobs—was felt almost as soon as the Grand Banks cod moratorium was announced on the East Coast in 1992. West Coast fishermen, many deep in debt, have every reason to fear that wild Pacific salmon stocks are going the way of the Grand Banks cod.

In 1992, a report by Carl Walters, fisheries professor at the University of British Columbia, said the wild coho population in the Georgia Strait was about 500,000, or less than half of what it was in the mid-1970s. The catch of wild coho from streams and rivers flowing into the strait is falling at a pace that, according to DFO itself, could lead to the extinction of these wild coho runs within the next thirty years.

The decline in wild coho—mainstay of the province's lucrative sport fishery—has been accompanied by the use of hatchery fish in the

Strait, leading to the suspicion that the hatcheries are actually hurting the situation.

"Hatcheries don't work," says Walters. "I pray they will all be gone and replaced by small projects, especially those that rehabilitate habitat. That, and better fisheries management."

Fisheries bureaucrats, prodded by critics, are moving to determine whether hatcheries are wiping out the fish stocks they were supposed to save.

DFO officials are not convinced by critics' research and point, instead, to the hatcheries' successes. A reliable chinook sport fishery in Port Alberni—which now calls itself the salmon capital of the world—is the direct result of nearby Robertson Creek Hatchery.

In 1977, Ottawa and Victoria (the province of B.C. is a minor player here, contributing no actual funds but simply technical support) unveiled the Salmon Enhancement Program (SEP) with a goal of doubling Pacific salmon runs and preserving valuable native, sport, and commercial fisheries. To date, this target has not been met and is not likely to be.

Twenty major hatcheries were built in B.C., primarily on the coast. Some became major tourist attractions, environmental showcases meant to demonstrate humans harmonizing with prehistoric forces and lending nature a helping hand.

The hatcheries turned loose millions of young salmon to bolster stocks and compensate for overfishing and habitat damage, at a cost of $20 million a year, half of SEP's annual budget.

At its inception, SEP mirrored the Zeitgeist of a time when there was a greater faith in science: Fish were to be manufactured as needed. Indeed, former fisheries minister Tom Siddon, in private life an aeronautics engineer, set the tone when he vowed that rather than reducing the salmon fleet, he would produce more fish.

SEP's high-tech empire was driven by engineers. In the early days, when bigger was always better, fishermen were seldom asked to forgo catches because more juvenile salmon could simply be manufactured and released.

Critics charge this solution to overfishing was, and still is, a poor substitute for responsible fisheries management.

As far back as 1982, the Pearse Royal Commission into the Pacific fishery warned that SEP's hatchery program posed a threat to wild stocks: "Probably the most widespread concern is whether artificially enhanced stocks will result in the destruction of natural stocks, frustrating the apparent gains by simply replacing wild stocks with enhanced stocks."

Twenty-six years later, still a vigilant observer of fisheries policy, resource economist Peter Pearse says B.C. fishing policies are often determined by political expedience rather than by the requirements of complex marine ecosystems. "Ministers often find it difficult to cut back on fishermen's catches to protect stocks from overfishing. It's easier to disguise the problem by cranking out more artificial fish from hatcheries."

In order to offer Vancouver's Expo '86 visitors the fishing experience of a lifetime, SEP cranked up its coho hatchery capacity and released ten million juveniles. Trouble was, the fish didn't co-operate and many simply disappeared in what scientists refer to as the "black box" of the Pacific Ocean.

Since 1986, the recreational fishery in the Strait of Georgia has collapsed, paralleling the decline of Chinook and coho salmon stocks. Coho, long the bread and butter quarry for sport anglers, has suffered an astronomical decline with a catch of more than one million in 1988 to about 10,000 a year by the turn of the century. Chinook catches have dropped from 200,000 a year to about 50,000 during this same time period.

Walters, who has studied the reduction of already threatened wild stocks when hundreds of thousands of artificial coho are pumped into the same feeding grounds, challenges conventional explanations like overfishing and habitat loss.

Using data collected by federal fisheries, he says declining marine survival rates of wild coho can only be caused by one of two things: dramatic changes in ocean conditions like El Niño's warm water current, or, more likely, intense competition between hatchery and wild coho for food and habitat.

According to Walters, wild coho compete with their artificial counterparts for limited amounts of herring, crab larvae, and krill. The result, he suggests, is the death at sea of increasing numbers of both wild and hatchery fish.

He contends that similar poor fisheries management and an overdependence on hatcheries brought commercial fishermen in Washington and Oregon to virtual ruin. Hatcheries were unable to replace wild stocks, and U.S. fisheries managers have ordered the curtailment of fishing in order to protect wild salmon.

For years, Walters has been calling for a rigorous and independent review of the interaction between wild and hatchery coho in the Georgia Strait. "While SEP is probably the most reviewed federal program in the country, somehow these reviews are always done by friends of the program."

There is a growing consensus that something has to be done to save wild salmon stocks. Warns Pearse: "Unless we get more wild stocks back to spawn, we may repeat the tragedy of the Grand Banks cod. The surest, most environmentally friendly, and by far the most economical way to rebuild salmon stocks is to let more wild fish reach their spawning grounds. That means catching less. It's as simple as that."

In 1994, Pearse conducted an economic analysis of all the works constructed under the SEP program, which revealed that costs far exceeded the benefits, and in many cases increases in enhanced fish were offset by decreased numbers of wild fish.

"The Salmon Enhancement Program was launched with great fanfare in 1977 with assurances from DFO officials that salmon stocks were declining, that they could not be restored through better management of wild stocks, that the technology of enhancement was proven and could be used to double salmon production, and that fishers would pay for it. After spending millions of taxpayers' money, all of these were proven wrong," Pearse wrote.

The impact on British Columbia's steelhead populations has been called genocidal. In river after river on Vancouver Island and on the mainland, steelhead populations are heading for extinction. The reason is not only hatcheries; destruction of river habitat by irresponsible logging plays its part. But hatcheries are delivering the final blow.

As former DFO minister David Anderson wrote in the *Globe and Mail*: "We are quick to criticize Europeans for their catches of Atlantic cod. It is ironic that we accept without protest government programs that are leading to the extinction of the greatest sport fish."

Just north of Vancouver, in Howe Sound, recreational angler Peter Hill has salmon on his mind as he bobs in the current, hunkered down in an eighteen-foot runabout. After tugging at his baseball cap he fiddles with a pink buzzbomb lure attached to the end of his trolling rod. Fishing is in his blood, a tradition for a man whose family leases a cabin at nearby Anvil Island, a thirty-minute spin across the water from Horseshoe Bay, the busy ferry terminal that connects the exploding population of the Lower Mainland with Vancouver Island and the Sunshine Coast.

Out in his boat, the world slows down; the insistent everyday pressures of modern life somehow dissipate, at least for a while. Hill never catches salmon, although he's been fishing these grounds for thirty years. But the summer of 2007 is different: wild pink salmon are returning in huge numbers, and he hears they will congregate off the south end of Anvil Island near Pam Rock, so he steers his boat in that direction.

As an English teacher at a Vancouver high school, Hill is up-to-date with issues relating to the politics of salmon, in a province where salmon have an iconic status. His attitude towards fishing has been tempered over the years. He wants to catch a salmon, yes, but he's not interested in wrecking the planet. As a child, fishing was all about the "catch-and-kill."

"We were hell bent on destroying this place," he recounts later. "As kids we would happily dump all our garbage overboard." And just around the corner, mills once spewed dioxins and mercury into the waters of Howe Sound. Then salmon fishing collapsed for a number of years. Now, there is hope some species will regenerate and come back.

When federal fisheries told us to reduce fishing, most of us listened, going to catch-and-release for species such as cod, which, according to Hill, are now much smaller in size. "I still love the cod," he says. "They've always schooled near Anvil and they put up a great fight."

An amateur historian, Hill has time, out on the water, to reflect on the clash of cultures as white settlers moved here on what was then Indian land. He well understands that our mostly European ancestors were not given this land, this ocean, as a gift: they conquered it in the great land grab, sweeping aside the aboriginal peoples, to take possession of rich renewable resources, salmon and timber pre-eminently—and set out to use it up as fast as possible. The lumber barons and fishing families made their fortunes not far from here.

In late summer of 2007, Hill found the pink salmon of his dreams. In one hour, he caught seven. And because the limit was four, he released three back. Hill was right. It was a good year for wild pink, the second of its two-year life cycle. An estimated eleven million returned, according to the Pacific Salmon Commission, even though the preseason estimate was twenty million. And the fish came back early, raising a caution for some experts: is this aberrant behaviour a threat to the species?

"After a lifetime of frustration and getting skunked, it was a rush to pull them in. However, I have to admit it got a bit boring to haul in numbers six and seven—a bit of a waste, really."

He also, by accident, caught and hauled a rock cod into his boat. Although Hill quickly released the shiny brown fish back into the water, he worried it would die anyway, its lungs bursting as it was hauled to the surface. Rock cod and other local groundfish have flourished in these deep waters for thousands of years, often living to the age of one hundred or more.

"Atlantic cod and Pacific salmon are quite different fish and the fisheries they support differ as well, but the management challenges they present are basically similar: they both call for clear policy, good science and strong management. And, in both cases, these have been lacking," Peter Pearse says.

Today, Pacific coast populations of rock fish too are diminishing, and one has to wonder whether extinction of one fish species after another will be inevitable. The Grand Banks cod collapse of 1992 may have been a warning shot, but how many of us truly heard it?

10

Offshore Oil Is the New Cod

"We'll pay for any good fortune that comes our way."
—Newfoundland saying

On October 8, 2007, Danny Williams and his Progressive Conservatives won a landslide victory in Newfoundland and Labrador. A buoyant Williams told the CBC he was "truly, truly overwhelmed by what has happened here tonight, " and added he was humbled by the magnitude of the victory. At the end of the night, Williams's Tories had taken forty-three seats, the Liberals three and the NDP one. Williams's vote share was the highest since the 1949 election, when Joseph R. Smallwood's Liberals formed the first post-Confederation government with 70 per cent.

Williams, who campaigned on a platform that emphasized hope and prosperity, rallied voters to back his agenda of greater control over natural resources and a plan to use oil-based wealth to improve an array of social services. The premier told supporters that Newfoundland and Labrador, which has been benefiting from high oil prices, would become a "have" province within two years. "We will be self-reliant," he told CBC Radio, echoing one of his campaign slogans.

That reliance will come directly from offshore oil and gas royalties. For more than a year before the election, Williams adopted an in-your-face negotiating style to successfully broker a deal to develop the proposed $5-billion Hebron offshore energy project. The agreement, which would extend the boom in the province's offshore oil industry, followed a protracted battle between Williams and a consortium of energy companies he derisively dubbed "Big Oil."

To date, Williams has built his successes on tough talk and a singular talent for the well-chosen fight. Still, the question remains: how long before a citizenry, sniffing the air and clamouring, and intoxicated by dreams for economic salvation—seemingly untempered by any notion of realism—turns against Williams, if the sum of the oil and gas largesse is not seen in the outports? Many hands will be looking for that oil money, needed to address a multitude of woes: shrinking communities, health care, and aging, diminishing economic opportunities, and fewer young families to take the reins of municipal governments, charities, and volunteer work.

Few in this province view the promise of wealth—offshore or otherwise—without a hefty dose of reality. On the Rock, the tendency is towards a kind of optimistic fatalism: "We'll pay for any good fortune that comes our way"—a common phrase, usually used to describe good weather: we're going to pay for that.

The story goes back to 1979. In that year the Hibernia P-15 discovery well was drilled by a feisty Brian Peckford, premier of Newfoundland, and Joe Clark, would-be prime minister of Canada for another five months. But by the spring of 1980, Pierre Trudeau was once again prime minister and on a constitutional mission that would not include sharing jurisdiction of the offshore oil and gas resources with Newfoundland.

Three more offshore oilfields would be discovered in the next few years—Terra Nova, White Rose, and Hebron-Ben Nevis. Off the Labrador coast, a group of companies led by Petro-Canada had also made five natural gas discoveries. Oil and gas wealth was just around the corner for Newfoundland, or so many people believed.

The first half of the eighties was marked by a nasty jurisdictional dispute over who owned the offshore oil and gas resources beneath the seabed: Newfoundland or Ottawa? The federal government offered an advisory role for the province in running offshore matters, but Newfoundland insisted on joint management. Both sides were flexing their muscles in setting regulations governing offshore permits, drilling, and hiring practices. The federal government launched the National Energy Program; Newfoundland created its own Petroleum Directorate to govern offshore oil and gas.

Many people in the province held the view that since Newfoundland had brought the Grand Banks and its fishery into Confederation with Canada, the province was the rightful owner of the resources above and below the seabed. That concept had not been enshrined in the Terms of Union—not for the fishery and not for seabed resources.

Control of resources is a recurring theme in Newfoundland politics. A resource-based economy, the province's fortunes rise and fall with commodity prices, new projects, and the next big find. A lack of control over how those resources are developed is viewed as a big reason why Newfoundland never seems to get rich from those resources the way Alberta has with oil. The Athabasca oil sands are primarily located in and around the city of Fort McMurray, which was still, in the late 1950s, primarily a wilderness outpost of a few hundred people whose main economic activities included fur trapping and salt mining. Since the energy crisis of the 1970s, Fort McMurray has been transformed into a boomtown of 80,000 people struggling to provide services and housing for migrant workers, many of them from Eastern Canada, especially Newfoundland. Alberta has become the white-hot centre of the Canadian economy, where riches are pulled out of the yawning black pits like cash from an ATM. The unemployment rate there is 3 per cent and salaries have gone through the roof. Inexperienced truck drivers make $100,000 a year; experienced welders double that amount. As a result the biggest hurdle facing companies, such as Shell Oil, is to find and keep skilled workers. "We're hoping these workers will go home to Prince Edward Island and Newfoundland and tell their friends that Shell is where they should come if they want to work in the oil sands," Shell spokesperson Janet Annesley told the *Wall Street Journal* on December 5, 2007.

In Newfoundland, news like that can reignite the pain of ancient injuries. Recent developments have only reinforced that notion:

- A depleted northern cod resource under Ottawa's watch.
- Labrador iron ore shipped out of the province without local processing.
- Hydroelectricity from the Upper Churchill has generated $19 billion in revenue for Quebec since 1960s *but only* $1 billion for Newfoundland, thanks to a deal that can't be renegotiated until 2041.

The federal government also takes some of the heat for the lopsided Upper Churchill contract. Rather than broker an agreement between two provinces, the feds were seen as siding with Quebec when the Newfoundland government asked for Ottawa's help in running transmission lines through that province to export Labrador hydro power direct to markets. Ottawa advised Newfoundland to find another way around the problem.

The other flaw in the Upper Churchill deal became apparent when electricity prices jumped in the early 1970s—the lack of an escalator clause to factor in rising prices. Under the agreement, Newfoundland continues to sell hydro to Quebec at 1960s prices, while Hydro-Quebec sells the same electricity to its American customers at current market rates. Other premiers have tried to undo the contract—issuing threats to turn off the power, going to the court, or using the power of persuasion and public opinion. To no avail; a contract is a contract.

The hydro contract factors big in Premier Danny Williams's recently released energy plan. It focuses on developing non-renewable resources, such as oil and gas, over the next few decades to help develop renewable resources, such as electricity. The entire plan works towards a deadline of 2041, not coincidentally the year the Upper Churchill deal expires. It also promotes the development of Lower Churchill hydro power.

By February 1982, the Peckford government referred the question of offshore ownership to the Newfoundland Supreme Court. The following month, the premier called a provincial election on the offshore issue, promoting offshore oil and gas development in Newfoundland and Labrador that led to the Hibernia Statement of Principles. On April 6, Peckford's Tories landed 61 per cent of the popular vote and forty-four of fifty-two seats in the House of Assembly. They claimed they had a mandate to press for control of the offshore.

In May of the same year, the federal government referred the same question—who owns the seabed resources on the continental shelf?—to the Supreme Court of Canada. Peckford said he was "shocked beyond comprehension by the arrogant and cowardly act" and proclaimed a day of mourning in Newfoundland.

The premier also condemned the federal government for referring the question to a higher court even before the Newfoundland Supreme Court had a chance to make its own ruling. In a pamphlet sent to every household in the province, the Peckford government asked this question: "Why are we being treated differently than other Canadians once again?" It resonated in a province that often sees itself as being treated like a poor, unwanted cousin by other parts of the country.

(Today, the Internet has merely given bloggers a new forum for this discontent. They don't have the market cornered on anti-Ottawa talk in Newfoundland. You can hear similar bitter views from the average "open-line" radio caller or Newfoundland nationalist in a local bar. The province has an abundance of them, and local politicians have made careers out of fighting with the feds. The federal government, in turn, has raised hackles with its tendency to ignore rural, far-flung parts of the country and their tiny electorates.)

Throughout the offshore dispute, Peckford professed his desire to set aside the ownership question and return to the negotiating table to hammer out a joint-management agreement that would allow Newfoundland to reap the bulk of oil revenues until it was no longer a have-not province. It's a theme Premier Danny Williams repeatedly hammers away at in his fight for oil dollars and equalization.

By 1984, the Supreme Court of Canada ruled seabed resources did belong to the federal government. That fall, the Progressive Conservatives won the federal election and Brian Mulroney was elected prime minister—something the Peckford government had been waiting for. The federal Conservatives offered a joint federal-provincial regulatory board with an independent chairman, offshore royalties for the province, and the usual taxes for the federal government. In 1985, the Atlantic Accord created the Canada-Newfoundland Offshore Petroleum Board to regulate the offshore industry from its offices in St. John's.

That same year, Mobil Oil announced it would develop the Hibernia oilfield, located on the Grand Banks about 315 kilometres southeast of St. John's, using a concrete gravity base structure rather than a floating steel platform. The company had spent six years studying its options, and had been heavily pressured by the province to opt for the GBS, which would provide job opportunities to unemployed Newfoundlanders. The provincial government naturally welcomed the decision. "What must be remembered is that a floating production system could be constructed anywhere in the world and floated to Hibernia without any construction activity occurring in Newfoundland," said Premier Brian Peckford in an *Evening Telegram* article.

The Peckford government also wanted what it considered to be a safer, fixed concrete platform operating in the same part of the North Atlantic where the *Ocean Ranger* oil rig capsized in a terrible storm, killing all eighty-four crewmen. The area is also known as iceberg alley—the route that icebergs calving off Greenland follow as they drift southward through the Grand Banks.

The Hibernia GBS is a massive concrete pedestal designed to withstand the impact of an iceberg. It was designed to survive a collision with a one-million-tonne iceberg without damage, and a six-million-tonne iceberg with damage that could be repaired. The latter was considered to be the largest iceberg the Hibernia platform might encounter.

The GBS was also designed to store 1.3 million barrels of crude oil that would be pumped aboard shuttle tankers to transport the crude to a transshipment terminal at Whiffen Head, Placentia Bay. From there, larger tankers would carry the crude to refineries in Eastern Canada and the U.S. eastern seaboard. (While the Terra Nova oilfield would use this facility to store its crude, White Rose uses larger tankers to ship its oil directly to market.)

The late eighties were marked by negotiations to get the costly Hibernia project off the ground, despite low world crude prices that dropped by half in 1986 from their previous highs of $34 US per barrel high. To get the project off the ground, the federal government contributed $1 billion in cash and $1.7 billion in loan guarantees. By September 1990, the royalty regime for Hibernia was finalized and construction of the Bull Arm site would begin.

On July 2, 1992, the day the cod moratorium was announced, about 1,000 construction workers were busy building the Hibernia gravity base structure in Bull Arm, Trinity Bay. Construction of the 600,000-tonne, concrete GBS was something of an economic lifeline in a province facing the layoff of 20,000 fishermen and plant workers from the northern cod fishery. Federal compensation programs would help fishermen and plant workers, while Hibernia poured construction dollars into the province.

It was a turbulent start for Newfoundland's first offshore oil project. When the cod stocks collapsed, the province's offshore oil industry was still in its infancy.

Six months earlier, Gulf Canada Resources had announced it was pulling out of the $5.2-billion Hibernia project, saying declining oil prices made it too costly to produce the crude. Just ten days after Gulf's announcement on February 4, 1992, and one day shy of the tenth anniversary of the Ocean Ranger sinking, the Hibernia partners announced layoffs and spending cuts at the Bull Arm construction site. Those cuts included:

- 450 layoffs at Bull Arm immediately and 200 layoffs later;
- 120 layoffs in Montreal and Paris;
- spending was cut in half—from $3 million per day to $1.5 million per day;
- ongoing work would continue with 1,000 people, but no new concrete would be poured;
- the startup date would be pushed back from 1996 to 1997.

When the cod fishery was shut down, Hibernia was already in a scaled-back construction mode.

Meanwhile, Gulf continued to offer its 25-per cent stake in Hibernia to dozens of other companies. There were no takers. The remaining partners—Mobil Oil, Chevron Canada, and Petro-Canada—fared little better in their search for new partners. Texaco, rumoured to be in the running to buy the Gulf stake, announced in January 1993 it wasn't joining the project, and the rest of the Hibernia partners were considering cutting their losses and walking away. By March, good news finally arrived.

The federal government would take an 8.5-per cent ownership stake. Murphy Oil bought 6.5-per cent and Mobil and Chevron took 5 per cent each. Ottawa would spend $431 million on the project over the next four years, and by 2002 it had earned that money back. Since then, the project has been earning profits for the federal government.

The delay in finding a partner had taken its toll on the project. By 1994, Hibernia was a year behind schedule and about $800,000 over budget. Finally, in February 1997, the modules were installed on top of the concrete GBS at Bull Arm. By June, the production platform began its 340-kilometre journey, under tow, to the Hibernia oilfield and was installed on the seabed.

On November 17, 1997, the $5.8-billion Hibernia project was reported to have "gushed" its first oil. To date, it is estimated that Hibernia has pumped 57 per cent of its oil reserves from the field.

In 2006, the Hibernia partners submitted an application to develop an additional 223 million barrels of oil in the southern part of the field. Their proposal called for drilling in the southern extension as early as 2008 and extended the twenty-five-year life of the main field by eight years.

The development plan was approved with several conditions by the CNLOPB (Canada-Newfoundland and Labrador Offshore Petroleum Board). In January 2007, the provincial government rejected the proposal. It said the plan didn't contain sufficient information about: the partnership, which is not the same as the original part of the field (for instance, the federal government isn't a partner); how the partners plan to develop that oil; or the kind of modifications that are required for the GBS to develop the oil.

Since both federal and provincial governments must approve any development plan, the Hibernia South extension is stalled until the partners submit a new and more detailed application. As of September 2007, they had not done so.

Compared with Hibernia's rough ride, the startup of the Terra Nova and White Rose projects were considerably less complicated, although not without expensive and time-consuming challenges. These smaller fields were developed more quickly, for less money and paid higher royalties to the province faster than Hibernia.

Terra Nova would be developed using a one-of-a-kind FPSO—a floating production, storage, and offloading vessel commonly used in the North Sea to develop oil. But this sea-going FPSO would also be ice-strengthened and designed to quick-disconnect from wells on the seabed in case of an emergency. The main emergency envisioned was an approaching iceberg that offshore support ships would be unable to either tow away or use water-cannons to shift away from the production ship.

In the late 1990s, Petro-Canada caused an uproar when it opted to keep its engineering team in Leatherhead, England, rather than move them to Newfoundland, where the company was building the topsides for the FPSO. (The hull was built in a Korean shipyard.) It was viewed as breaking a technology transfer promise. Under its development plan approved by the CNLOPB, the company was supposed to move as many as forty-five engineers (many of whom were Newfoundlanders) to the province "as soon as practicable" after the project received the go-ahead.

The moving date was supposed to be sometime in the fourth quarter of 1998. As that deadline approached, Petro-Canada argued that doing so would adversely affect the project's schedule and budget. In June 1998, the CNLOPB ruled the engineers could remain where they were in England.

The project wound up delayed and over-budget anyway—but for different reasons. In February 2001, Petro-Canada announced the $2.5-billion Terra Nova project was 15 per cent over budget. That pushed the price tag for the project in to the $2.8-billion range. The company also pushed back the scheduled date for first oil for a third time (the new date was Q4 2001 and the previous target was June of that year.) Some of the work done overseas had to be completed at Bull Arm, where the FPSO hull and its topsides would be joined together.

Petro-Canada has admitted that some budget decisions early in the project would come back to haunt them. It said the decisions were made when oil was $12 US per barrel and keeping project costs low was a major consideration. The company also pointed out Terra Nova's low capital costs allowed it to reach payout faster—pumping 30-per cent royalties into provincial coffers in the process.

On November 21, 2004, the FPSO spilled up to 1,000 barrels of oil into the Atlantic Ocean. The spill was attributed to a faulty valve in a unit that separates oil, gas, and water. (The oilfield resumed production in about five weeks.)

In December 2004, the CNLOPB ordered Petro-Canada to clear up a backlog of maintenance work that it said was unrelated to the oil spill the previous month. The FPSO was also experiencing mechanical problems with its gas-compression system. Without it, the FPSO can only pump oil at reduced rates. Some of these problems were related to the earlier budget decisions.

By the fall of 2005, Petro-Canada was considering how to carry out a bevy of maintenance repairs during the summer of 2006—and one option was dry-docking the FPSO at a European shipyard. The plan called for heading to Rotterdam in July for a ninety-day refit. The field shut down earlier than planned when the second of two generators aboard the FPSO failed in May.

The Terra Nova refit lasted six months and cost $225 million. The shutdown also took a $226-million bite out of provincial royalty revenue during that budget year. The FPSO returned to the field in September 2006 and reconnected to the seabed the following month. By 2007, it was producing at normal production rates.

The $2.35-billion White Rose project is the third offshore project. With reserves up to 250 million barrels of oil, it's also the smallest of the offshore fields—a point that Husky Energy would emphasize in its choice of production platforms. The company also preferred using a less expensive FPSO, just as Petro-Canada had done. Husky filed its development application in January 2001.

During the public hearings on the White Rose project later that year, an organization called Friends of Gas Onshore—FOGO, for short— pushed hard to persuade Husky to build a Hibernia-style gravity base structure. The group envisioned such a GBS as a hub to develop natural gas throughout the Jeanne d'Arc Basin—combining the gas resources of Hibernia, Terra Nova, and White Rose.

That debate also hinged on whether or not White Rose contained natural gas in sufficient volume to be economical to be produced. Husky said the field contained 1.8 trillion cubic feet of gas, while the CNLOPB said it held 2.7 tcf. Either way, it wasn't enough for a gas project. And the company maintained more delineation wells were needed to prove these resources.

The other issue: none of the natural gas at the other two fields was under development. Some gas is used to generate electricity aboard the Hibernia platform and the Terra Nova FPSO. The rest is re-injected into production wells to stimulate the flow of oil. White Rose also uses some of its gas and stores the rest in another part of the field.

The Husky proposal to use an FPSO was approved, and the company was required to submit quarterly employment and spending reports to the CNLOPB—outlining its efforts in achieving Newfoundland–Canada content. The FPSO hull was built in Korea, as was the hull for the Terra Nova field.

Unlike Terra Nova, all the engineering and design work for the topsides modules was done in Newfoundland. Seventeen modules, most of them locally built, were installed aboard the FPSO hull at the Cow Head fabrication yard, which is part of the Marystown Shipyard on the Burin Peninsula. The facility is owned by Peter Kiewit Sons. The Marystown workforce peaked at 1,400 during the White Rose project.

By the summer of 2005, the completed FPSO was preparing to sail to the oilfield. During the official blessing of the production ship in

Marystown, Premier Danny Williams called White Rose a model for future offshore developments—one that included all engineering and technical work done in the province: "The White Rose approach contributed significantly to technology transfer and the further enhancement of local expertise in this industry. Of all the developments in the province to date, I firmly believe the White Rose project most accurately represents our future path," he said.

Williams also said the project had raised the bar for future developments—and he expected future development to surpass the White Rose benefits. The Hebron partners may have assumed the premier was talking about local engineering and construction work.

Williams was aiming higher. By February 2006, he would outline his three additional requirements for the next offshore oil project—equity, a petroleum refinery, and super-royalties when world crude prices are high. And he was willing to settle for two out of the three requirements: "The irony is, here we are with the resources adjacent to our province—we brought them into Confederation—and we don't have an interest in our own resources," he was quoted as saying in a February 10, 2006 article in the *Evening Telegram*.

"I want the people of Newfoundland and Labrador to have ownership in these projects. We're not looking for the moon on ownership here, but we want ownership."

In summer 2007, Husky received a green light to develop an additional 24 million barrels of oil in the southern part of the field. That extension will cost $595 million to develop. It's the first of three planned satellite expansions that will almost double the size of the White Rose oilfield. Combined, the three satellites contain about 214 million barrels of oil. That could increase as Husky drills more wells to further quantify oil reserves there. Husky is aiming to produce oil from one of the satellites by the end of 2010.

As part of the expansion, Husky also has a new partner in the Newfoundland government. Premier Danny Williams announced, on September 12, 2007, that the province had bought a 5 per cent ownership stake in the White Rose satellite fields for $44 million Canadian, based on current reservoir information. The province's energy corporation, which is a subsidiary of Newfoundland and Labrador

Hydro, will also pay Husky a $3.50 processing fee for every barrel it produces.

As well, Husky will pay super-royalties of an additional 6.5 per cent whenever world prices for West Texas Intermediate crude are above $50 US. That royalty applies when the project reaches net payout—meaning when the costs of developing the field are paid off, plus a return allowance for the companies. The royalty regime for the original part of the White Rose field remains the same. The provincial government estimates it will reap more than $6 billion in revenue from the satellite developments.

Estimated to cost up to $5 billion to develop, the Hebron project is on again. In June 2007, at an offshore oil and gas conference in St. John's, Premier Williams announced that informal talks had begun on the Hebron. By August, the tentative deal was announced. The Hebron partners led by Chevron Canada and the provincial government reached a memorandum of understanding (MOU) to develop the field. A final agreement is expected to take another eighteen months or so. Premier Danny Williams announced the memorandum of understanding on August 22—just thirteen days after the premier's office had confirmed formal negotiations had resumed.

While full details of the MOU have not been released, as negotiations to reach a final Hebron agreement continue, the province will get a 4.9-per cent equity stake in the oilfield for $110 million up front, and its portion of development costs in future.

The province will also receive super-royalties when crude prices top $50 US. In return, the project will get a break on royalties in its early production years until Hebron reaches payout—the point at which development costs of the field are recouped.

Why is equity so important? It's all about control.

Newfoundland governments often believe they don't have any control over the big projects and resources developments in the province. In the past, Newfoundlanders have been all too willing to give up control: of a country by joining another country; of the fishery when joining Canada; and of an elected government that was willingly given up and replaced by an appointed commission of government when the country of Newfoundland faced bankruptcy in the 1930s. Slowly, a sense of outrage is being replaced by a growing confidence and an unwillingness to be easily dismissed because of how small the province is or how far removed it might be from the centre of power.

Not every Newfoundlander may have cared for Premier Williams yanking down the Canadian flag during the Atlantic Accord talks with the Martin government, but they seemed to like the fact that the issue wasn't ignored.

His stated goal is to model Newfoundland's petroleum development on that of Norway, where state-owned oil and gas companies lead or are partners in offshore oil and gas projects. It includes developing a level of engineering and manufacturing expertise in offshore projects that can be exported around the world—much as the Norwegians and their companies did when the Hibernia project was under construction.

Williams also wants a piece of the resource ownership pie. He swept into power campaigning on an energy plan that calls for a provincial equity stake of 10 per cent in all future oil and gas projects off Newfoundland and Labrador. That idea isn't a hard sell in Newfoundland where the perennial complaint is that others usually reap the rewards of the province's resources and all locals get is a few construction jobs. Chevron is reassessing the cost of developing Hebron. The price tag will likely be higher than the original $5 billion, given the rising costs of both labour and material. The company also said first oil for Hebron is anywhere from eight to eleven years away.

The offshore oil industry is transforming the economy of Newfoundland and Labrador—but the bulk of the direct impact has been felt on the Avalon Peninsula, where almost half the province's population lives. That is where many of the direct jobs, indirect jobs from spin-off activity servicing the offshore, and the bulk of the offshore infrastructure are located.

St. John's Harbour and Bay Bull Harbour are home to offshore supply bases for the industry. Newdock located near St. John's Harbour manufactures seabed equipment for the offshore. Industrial parks in St. John's, Mount Pearl, and Paradise house offshore service and supply companies—providing lay-down yards for steel pipe used to line wells that are drilled offshore, fabricate offshore equipment, or warehouse other supplies.

Bull Arm, built for the construction of the Hibernia GBS, has also been used, to a lesser extent, to build and test the wellhead equipment that sits on the seabed at the White Rose project and to build steel topsides modules for both Terra Nova and White Rose.

The Marystown shipyard, once a waning facility operated by the Newfoundland government, was refurbished and modernized by Peter Kiewit Sons. The yard and its Cow Head fabrication facility built some of the modules and installed the topsides aboard the White Rose FPSO.

How much of an impact the offshore oil industry has had perhaps can be measured against traditional resources sourced in Newfoundland. It is estimated that offshore oil accounted for 15 per cent of the province's real GDP in 2005, according to *The Economy for 2006-07*—an annual provincial publication released with the provincial budget.

Combined, mining and oil extraction accounted for 18.7 per cent of the provincial GDP. The province's gross domestic product in 2006 stood at more than $14.08 billion. The 15 per cent figure cited suggests oil accounts for the bulk of the 18.7 per cent GDP figure. For the most part, that number has grown steadily since 1999. While the fishery employs more people, its state of decline remains obvious: GDP numbers from 2006 indicate:

- 2.57 per cent—agriculture, forestry, fishing and hunting
- 6.45 per cent—manufacturing, which includes fish products
- 1.27 per cent—seafood product prep and packaging
- 14.53 per cent—finance, insurance, real estate and business support services
- 8.28 per cent—public administration (government)

In 2006 and 2007, the three producing offshore oilfields spent more than $1.34 billion on their operations in Newfoundland and Labrador. During the same year, they directly employed 2,929 people.

All production facilities and oil rigs have a 500-metre exclusion zone in which no fishing boats or other vessels may venture, according to the Canada-Newfoundland and Labrador Offshore Petroleum Board. That 500-metre zone extends from any equipment, such as pipes or wellheads, on the seabed or anchor chains. Those zones are included on navigation charts for the Grand Banks area.

Fishermen and other mariners are also notified whenever seismic surveys are being conducted over offshore fields. As for the impact of those seismic surveys on fish and fish habitat, the few studies that have been done to date have been inconclusive. Fish in the immediate area of an air-gun blast are startled and change swimming directions and behaviour, but appear to recover within minutes or hours. Long-term effects may be possible, but more study is needed to determine this.

One study of crab, off Cape Breton, suggested serious effects on crab larvae, though not necessarily on more mature shellfish. Based on the limited number of studies that have been conducted to date, there is considered to be a high probability that some fish within the general vicinity (i.e. hundreds of metres) of a seismic survey operation will exhibit startle responses, changes in swimming speed or direction, and changes in vertical distribution, with recovery likely within minutes to hours after exposure.

There is a lower but still reasonable probability that seismic surveys will influence the horizontal distribution and catchability of some fish under certain conditions, such as during the migration of pelagic fish. If horizontal dispersion does occur, impacts are more likely to be observed over greater distances (kilometres) and for a longer duration (days). Seismic surveys are considered unlikely to result in immediate mortality of fish; however, sublethal physical damage and physiological impairments may occur within close proximity to an air-gun source and could potentially result in delayed mortality or chronic effects. However, additional research is required to assess the intensity of sound levels or typical ranges from a known seismic source required to produce these types of effects. The potential for seismic surveys to disrupt communication and other sound-dependant activities of fish is essentially unknown, as is the long-term ecological significance of the impacts described above.

Ironically, no-fish zones around the rigs may help to protect cod spawning and rearing grounds from the draggers.

Meanwhile, back in St. John's, as the premier continues to jab and joust with the prime minister, there is still no sign of a comprehensive Grand Banks cod recovery program. Nor, tellingly, was this subject discussed in Williams's victory speech.

Why not? A request for an interview with Premier Williams was summarily dismissed by a communications staffer in a BlackBerry email. Pressing on with this reasonable query: Why not, once a formal Grand Banks cod recovery program has been crafted, 1) use a small percentage of the offshore oil royalties to set it in place; 2) reduce the offshore dragger fleet with a buyback program accompanied by real-world re-training or education; and 3) turn the whole of the Grand Banks into a no-fish zone so the cod stocks can regenerate over a period of years, even decades? This may mean eliminating the shrimp fishery as well, as it bulldozes and destroys cod habitat.

Consider this: It could be Premier Williams's greatest legacy: A Grand Banks Heritage Fund, a Brave New Idea, generated and implemented in Newfoundland.

Political tensions between St. John's and Ottawa remain, of course, but, considering what the finality of the decimation of the cod says about us as human beings, there is reason to hope—assuming that Premier Williams and other politicians, the fishing industry, and the seafood companies can muster the will.

11

Peter Pearse, Rational Man, and the Prophet Daniel

"If we carry on life like this, the only things in the sea will be jellyfish and plantation soup."
—Daniel Pauly

"Fisheries around the world have to stop being simply harvesters and start being managers as well."
—Peter Pearse

Can Brave New Ideas bring the cod back to the Grand Banks? Was the federal Department of Fisheries and Oceans— derided as an agency that couldn't manage a home aquarium— responsible for developing and implementing policies that led directly to the collapse of the Grand Banks cod, one of the world's worst ecological disasters?

Yes, say two remarkable figures. Four thousand miles to the west of the Grand Banks, on a campus by the edge of the Pacific Ocean, two professors with starkly contrasting world views have thought long and hard about the collapse of the Grand Banks cod—and its implications for fisheries around the world.

Daniel Pauly

Biologist Daniel Pauly is a professor at the Fisheries Centre, University of British Columbia. Peter Pearse retired from the economics faculty at the same institution in 1996. Each knows well the other's ideas and reputation. But there is little warmth between the two men.

"Economists are not only wrong—they're stupid," Pauly says, relishing what he sees as the imminent death of fisheries economics. Mathematics has so dominated economic thinking that economists have become isolated from the real world.

"But he's just a biologist who wants to close down all the fisheries," counters Pearse. "Anyone could manage a closed fishery."

Innovators both, their ideas, fine-tuned during the past quarter century, are now discussed around the world, at a time of increased concern about overfishing in the wake of the Grand Banks cod collapse. Early in their academic careers, both realized they would need to reach out beyond the ivory tower to effect the sweeping structural change they envisioned, the change the world needed.

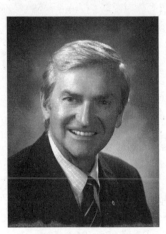

Peter Pearse

Both workaholics, with egos the size of the Sargasso Sea, they worked weekends to hone the intellectual tools needed to help transform a medieval catch-and-kill way of life into a sustainable, modern-day fishery that could feed fishing families and feed the world. To lay it bare for all to see: the greed, the folly, the waste, and destruction. To uncover the intellectual constructs that underpin a now bankrupt and heavily subsidized industry, one in which fishers in high-tech vessels bulldoze the ocean floor, and compete with each other to catch the last cod, tuna, or salmon.

To make an impact, both men realized early in their careers, they would have to reach out beyond the pedantry of academic publishing. To win the hearts and minds of a fickle public, they would make the media their weapon of choice.

Daniel Pauly, sixty-one, has become, by sheer force of personality, the world's pre-eminent biologist. The brightest star in the fisheries constellation, he has rightly been credited with blowing the lid off the world's overfishing. Googled in the millions and replayed on FaceBook and MSN, he is changing the way we look at fishing. In October 2007, he shared the stage with Al Gore, as he collected an award from Hollywood star Ted Danson's Oceana charity. Pauly and Gore were praised for their work in protecting the world's oceans.

Biologists, like environmentalists, tend to hold a gloomy world view. Crusader, evangelist, and scold, Pauly warns, in theatrical language, of an oceanic apocalypse, where, idled on a briny sea, we will sit down to a dinner of stewed jellyfish, another of his clever metaphors.

Building on the first great wave of environmentalism in the 1970s, Pauly conceptualized and popularized this and many other crossover phrases that now resonate in the Zeitgeist. "Fishing down the food chain" is another. It refers to our habit of first taking out the apex predators—large species such as cod, tuna, and swordfish—because they are the most desirable, then once they've been decimated, taking out their prey species (plankton eaters such as anchovies), and so on.

"We are moving from a marine ecosystem," wrote Pauly in a 2000 *American Scientist* article, "dominated by big fish to a soup of small organisms. If we carry on like this, the only things in the sea will be jellyfish and plankton soup."

"Ecosystem management" is another of Pauly's crossover phrases, meaning to see beyond an individual species, to see the ocean as a community of organisms, functionally linked with complex structure and biological diversity. For years, he's been urging governments around the world to set aside "no-fish zones" of marine-protected areas so that fish species could regenerate—so that spawners might, for a few years, escape the 100-metre-long weighted nets that plow the ocean floor like a giant bulldozer, shredding its habitat.

Pauly and his colleagues have also made a splash with paper after paper in the most prestigious scientific journals. The news is uniformly bad. And he's become a man on a mission: to spread the word that fish stocks are plummeting around the world. As he explained to a *New York Times* reporter:

> In some places in the world, you can see people chasing the last fish. In the Java Sea in Indonesia, I have seen fishers going out in the morning, six of them going out and coming back with five pounds of fish. That is the end point—a pound of fish per person per day to sell for rice. That's where fisheries go if you let it happen. That's where it stabilizes. These people cannot feed their families.
>
> We've declared war on fish—and we've won it. Fishermen now use military techniques to catch fish, employing tools like GPS, sonar, and satellite imaging to catch more efficiently. Our catches have increased fivefold between 1950 and 1990 . . . but we're also seeing a decline in fishing stock since the 1980s. Now 70 per cent of global fishing stocks are overexploited.

Unchecked, he says, the same will be seen around the world, and the fishing industry will leave little in the seas for future generations, but harvests of what he calls "bait and worse," the bottom levels of the marine food web, the sea cucumbers, jellyfish, and plankton.

The problem, according to Pauly, can't be remedied without large-scale government intervention: a huge reduction in global fishing and the radical step of creating "no-take zones," where fish can grow large, breed, and replenish. That won't happen, he maintains, unless the true owner of the ocean resources, the public, demands it, which has yet to happen.

Pauly has been alternately described by colleagues as inspiring, arrogant, brilliant, and aggravating. Some call him a heretic, "the Prophet Daniel." To others, his message has become increasingly shrill. Certain fisheries scientists have taken him to task, turning his data on its head. Professor Ray Hilborn of the University of Washington has posted a PowerPoint rebuttal on the Web. Hilborn also attacked the authors of a 2006 article in the prestigious magazine *Science*, which predicted the collapse of all of the world's fisheries by 2048, based on declining fish harvest numbers and other research. It also sparked a firestorm of controversy, generating headlines nationwide in newspapers and news magazines, spinning off into an elaborately illustrated feature in *Time* magazine. According to Hilborn, the article was "probably the most absurd prediction that's ever appeared in a scientific journal regarding fisheries."

Hilborn called the *Science* article findings "silly," but also worried that they "will become completely accepted in the ecological community. They have no skepticism." Hilborn and others refer to the growing number of similar reports as "an increasing Chicken Little response." The principal objection is that the scientists infer that fisheries are going to "collapse" based on declining catches. But one reason for the decline, he said, has been a successful management program. "The basic way they measure 'collapse' is flawed. Catch is not a good way to measure the status of the fish stock."

The authors of the original paper acknowledged that there is some validity to Hilborn's argument. "Yes, catches are an imperfect measure of the stock abundance," said lead author Boris Worm, a marine biologist at Canada's Dalhousie University. He added that declines in catches are still indicative of larger trends. "It's obvious that when the catches collapse, it's often because there's no more fish to be found."

Hilborn said many of the world's fisheries are not well managed and are getting worse, but the United States, Iceland, New Zealand, Australia, and others have successfully pursued strategies to keep fisheries sustainable. For instance, those countries are getting rid of a fishing industry race that led fishermen to build and operate ever-bigger boats to bring in ever-bigger catches. Lowering the take, he said, is the key. Hilborn describes himself as "an ardent conservationist," but said he worries that public exaggerations of environmental problems erode the credibility of scientists and the conservation movement.

Critics such as Hilborn are dismissed with a chuckle. "I understand there is one web site dedicated specifically to attacking my views," says Pauly in a followup interview. "But I haven't been to the site. Besides, I have far too much work on my plate."

Pauly is used to adversity. He struggled to find a place in the world where he belonged, a struggle that gave him the global perspective that has helped him become one of the most influential fisheries scientists in Canada.

He was born in 1946 in Paris to a white French mother and a black American father who abandoned the family. When he was two, Pauly went on what was supposed to be a short visit to the home of a Swiss family who had recently befriended his struggling young mother. But the family, Pauly recounts, refused to return him to his mother, telling him that she had abandoned him and, he learned later, they sent threatening letters to his mother in France.

Over the next fourteen years, he endured a bizarre Dickensian childhood. He was a forced replacement for the Swiss family's young son, who had died, all the while being turned into a live-in servant who cleaned and did other household chores. His identity crisis was compounded by being a half-black oddity in an all-white town. He found his solace in books.

At sixteen, he ran away and put himself through high school in Germany, eventually reuniting with his mother in France and with his father in the United States, where he learned the harsh politics and reality of race in 1960s America.

He returned to Germany, where he went to college and received his doctorate in fisheries biology at the University of Kiel. Eventually he travelled to western Africa to study fisheries, hoping that he might blend in better there. Instead, he says, "I found I was European."

Pauly sailed the Java Sea in Indonesia and carried out fisheries research in the Philippines, finally landing in Vancouver, a city of many immigrants from many cultures. In 1995, Pauly became a landed immigrant, retaining his French citizenship to this day. Easily identifying with the disadvantaged, Pauly has been working for years on "levelling the playing field," as he likes to say, trying to help scientists in developing countries conduct their research despite scant resources.

Most recently, Pauly helped create what may be the biggest and most lasting field-leveller of all, FishBase, online at www.fishbase.org, which is packed with information on every one of the world's 27,000 fish species, including photographs. It gets as many as five million hits a month. Still, giving the world access to information about fish is not enough for Pauly, who, critics and fans agree, has the gift of seeing the bigger picture.

For the past twenty-five years, Pauly has worked to put together a top-notch team of men and women who now work out of the new Fisheries Centre on the campus of the University of British Columbia. Recently, a film crew came to the centre to interview Pauly, who is one of the characters in what will soon be a "cinematic documentary thriller" about overfishing.

Like other high-profile greens, he benefits from the largesse of U.S. conservancies. An annual $1-mllion grant from the Pew Charitable Trust allows him to initiate new projects he would not be able to fund under UBC's current budget. Funding from this U.S. conservation foundation offers two other benefits: his work is featured and promoted by Pew officers, who post it prominently on the foundation's web site. It provides Pauly with an intellectual freedom that few professors today enjoy, a safe harbour for his intellectual explorations. Such freedom is not to be taken for granted, given the sordid and surreal aftermath of the Grand Banks cod collapse wherein politics so often trumped and even botched science.

He is a revving engine of a man, relentless and given to overwork, which led to a stroke in January 2004. Since then, he has recovered his gross motor functions—("I can hold things with my right hand which I use for typing.")—but has not yet recovered fine motor functions. "Have I 'recovered'? I guess 'partly' is the word," he explained in an email.

In his office on the campus by the sea, Pauly uses the back of an envelope to show how scientists in the late 1980s made serious, grievous errors when they attempted to set catch limits for the Grand Banks cod, using a model known as VPA (see chapter 3). As a result, industrial-scale assaults by the Canadian offshore dragger fleet hammered the stocks to near extinction.

Behind the assault, experts say, are steady advances in technology, national subsidies to fishing fleets, and booming markets for seafood. Demand is up partly because fish is considered healthier to eat than chicken and red meat.

Directed by precise sonar and navigation gear, more than 23,000 fishing vessels of over 100 tons and several million smaller ones are scouring the sea with trawls that sweep up bottom fish and shrimp; setting miles of lines and hooks baited for tuna, swordfish, and other big predators; and deploying other gear in a hunt for seafood in ever-deeper, more distant waters.

Flash freezers allow them to preserve their catch so they can sweep waters right to the fringes of Antarctica. The trade is so global that an eighty-year-old Patagonian toothfish hooked south of Australia can end up served by its more market-friendly name, Chilean sea bass, in a West Vancouver bistro.

The Grand Banks cod, which once could reach six feet in length, have essentially vanished. Despite closures of fishing grounds, they may never come back, Pauly and other biologists warn, because overfishing has so profoundly changed the ecosystem. Draggers with weighted nets ploughed the Grand Banks like bulldozers.

One consolation to biologists measuring such changes is knowing that a fishery is abandoned because of plummeting yields, so fish stocks still exist, although extinction might be near. The flip side is that as industrial fleets push into new waters, experts say, the danger and damage spread. The laws and international pacts that do exist can be circumvented, producing persistent illegal markets for coveted species.

The global fleets are sustaining harvests only by moving into untapped resources, says Pauly, who co-authored *In a Perfect Ocean*, a detailed analysis showing enormous drops in North Atlantic catches over the last century. "It is like a ring of fire burning through a piece of paper," he says. "Since the seventies, when the big fishing areas of the Northern Hemisphere saw catches drop, you've had this front moving out, with a massive effort off West Africa, in Southeast Asia, the southern Atlantic."

Scientists add, global fishing is spreading so fast that it is devastating marine ecosystems before scientists study them or get a rough idea of the size of populations. Off the coasts of North America and Australia, for example, biologists probing ridges and seamounts have found areas where trawls have uprooted communities of cold-water corals and other bottom dwellers that are centuries old.

Recent studies estimate that stocks of many fishes are now a tenth of what they were fifty years ago. As prized species have diminished, fleets have gone further down the food chain, for smaller fish, more squid, even jellyfish and shrimp-like krill. Industry calls it "biomass extraction" and turns the harvest into everything from fish sticks to protein concentrates for livestock or pellets to feed cage-raised salmon. International agreements protect some species, like tuna and swordfish in the Atlantic. Most fisheries in international waters are rarely monitored.

Falling catches have led to fast growth in fish farming and other aquaculture. These activities have exacted an ecological price as well. Salmon and shrimp farms expanding in coastal waters from the Bay of Bengal to the Bay of Fundy displace ecosystems that are nurseries for much sea life or threaten local species through releases of nutrient-loaded waste, non-native species, or diseases.

Experts say the industry expansion has been driven by growing populations and prosperity around the world. Almost a billion people now rely primarily on fish for protein.

Another factor is persistent subsidies that give fishing fleets breaks on fuel costs, vessel construction, insurance, or other expenses. All told, according to private analysts and the Food and Agriculture Organization of the United Nations, the subsidies amount to about $15 billion a year, or more than a quarter of the $55 billion in annual global trade in seafood.

Japan alone provides close to $3 billion in support for its fishing fleets. Support in the United States includes $150 million a year in tax rebates on marine diesel fuel, according to the World Resources Institute, a private research group. The subsidies are challenged by environmental groups and conservative organizations adopting free markets, including the Cato Institute. The problem, they all say, is simply that such aid results in too many boats for the available fish.

Another factor has been rapid advances in fishing technology. Much of the progress has been electronic: satellites of the global positioning system let fleets know their exact location, while increasingly sensitive and powerful sonar gear produces detailed readouts of schools and nooks where fish may lurk.

Ted Brockett, president of Sound Ocean Systems in Redmond, Washington, which makes and sells devices for ocean vessels, said technology could help stem fishing damage if fleets used the innovations not to pursue the last fish but to find the right fish—the size or species that can be harvested without degrading ecosystems. Long before then, ocean scientists and policy-makers say, the continuing fishing threatens to damage the ecological foundations of fisheries in ways that may last for generations.

In June 2007, the Pew Oceans Commission—with a nonpartisan membership including fishermen, scientists, and elected officials—recommended a "serious rethinking of ocean law, informed by a new ocean ethic."

But—and this is Pauly's takeaway message—the most important recovery strategy of all is simply to fish less. This can be accomplished in many ways.

Harvest limits can be set, with quotas allotted to individuals in a fishery, who can then trade them. Iceland has set the standard for this approach, which has also been adopted in a few American fisheries. By limiting the overall catch and allowing people to buy and sell their fishing rights, the system encourages some to leave the business. Environmental and conservation groups support this practice.

Fishing pressure can also be cut by creating marine reserves or closures that create nurseries. Pauly and others biologists have proposed that 20 per cent of the oceans be set aside. Reserves in coastal waters have already proved their worth, with rising catches in nearby areas. A notable success has been in St. Lucia, in the Caribbean, where reserves established in the mid-1990s increased nearby catches up to 90 per cent.

Some closures in United States waters have led to sharp recoveries. After a shutdown of bottom fishing in 1994 in New England, he said, scallops came back to record levels and overall abundance soared.

Today, Pauly and his team have called on the fishing nations of the world to end government subsidies of fishing fleets. Fish that live for more than one hundred years and cold-water coral reefs that have taken millennia to form are being destroyed by the "roving bandits" of the high seas that could not survive without government aid, Pauly points out.

Japan, South Korea, Russia, and Spain lead the list of nations that are sanctioning the pillage of the deep ocean with public money diverted into subsidies to pay for fuel and equip trawlers. Both Pauly and his Fisheries Centre colleague, Ussif Rashid Sumaila, present a devastating account of the effects of industrial deep-sea fishing on the Grand Banks. Once-abundant aquatic life has declined, they say, to the point where we probably have less than 5 per cent of the total mass of fish that once swam in Europe's seas.

This leads to another Pauly concept: shifting baselines, a term he coined in 1995 in his paper *Anecdotes and the Shifting Baseline Syndrome of Fisheries.* It refers to the loss of perception of change that occurs when each generation redefines "what is natural." With each passing generation, not just the number of fish in the sea but also the number of fish the public thinks should be in the sea diminishes. This phenomenon allows us to adjust to a depleted ocean without quite knowing what's slipping away.

Pauly developed the term in reference to fisheries management in which fisheries biologists sometimes fail to identify the correct "baseline" population size (how abundant a fish species population was before human exploitation) and thus are working with a "shifted baseline." Intensive fishing since medieval times has caused this decline gradually over the centuries, so that today's fish-deprived sea seems normal to modern generations. Now industrial fishing, especially trawling, has virtually eliminated entire habitats, including cod in Canada. Sophisticated devices such as sonar depth sensors are being used to plunder that last frontier—the deep sea.

Global positioning devices attached to floating logs alert fleets to the whereabouts of fish: Purse-seine boats now seed the ocean with veritable forests of floating decoy logs and other fish-aggregating devices to bring together scattered shoals of fish. When they return, they scoop up the fish with ruthless efficiency, taking with them turtles, sharks, and dolphins— or whatever happens to be in the wrong place at the wrong time. For some reason, logs preferentially attract juvenile tuna, so their take even of the target species is wasteful. By catching young tuna before they reach adulthood, purse-seine vessels forgo much higher catches for themselves later, and they are also denying these tuna the chance to reproduce, putting future catches at risk. Where once the vast canvas of the sea was great enough for fish to lose themselves in, escaping capture, today even the high seas afford little refuge. New technology has given old fishing methods a far more lethal edge.

Unless governments establish marine reserves in which fish stocks have a chance to recover, Pauly predicts that by the year 2048, fisheries for all the fish and shellfish species we exploit today will have collapsed. There are times, he says, when the capacity of mankind to blind itself to plain reality is simply breathtaking. Today, for example, he says, we still believe, as was universally believed two centuries ago, that the seas surrounding us afford an infinite source of wealth.

"A boundless delusion," Pauly says. "This isn't fishing anymore—it is the extermination of a species. And sadly, the fishing industry appears to have a far louder and more persuasive voice in the halls of government than do the fish themselves."

If Pauly is poke-in-the-eye, Peter Pearse is a rational man who happens to be the author's uncle. Although he too is capable of the thirty-second clip, he chooses diplomacy to sell his ideas to government and opinion leaders. Indeed, since retiring from UBC in 1996, he has had a successful second career as a consultant, travelling around the world to investigate and make recommendations relating to resource issues such as water, forests and, always, fisheries.

Today, mid-afternoon in late summer, sipping tea on an outdoor patio, the retired professor warms to a subject he first introduced to the Canadian government in 1979: individual quotas as a means to stop the "race to the fish." It was an idea that didn't take back then.

At seventy-five, but looking fifteen years younger, Pearse has the spare build, the leanness, the fitness of an older man who wants you to know he will probably outlive you. His Sidney Island home, set aesthetically into the stone bluff is leeside to the ocean, out of the prevailing winds. Perhaps his good health reflects the often sunny world view of economists: If only people would be reasonable, we could solve most of our problems—even the Grand Banks cod debacle.

There was no safe haven for Pearse for many years. He had the audacity to tell commercial fishermen in the early 1980s that they were part of a world economy, one propelled by venture capital and technological innovation. He warned we can never return to the village life of a pastoral civilization. The ghost towns of Newfoundland and the Pacific Coast are symbols of the devastation wrought by the ever-ratcheting effects of new technologies and industrial fishing.

Pearse's ideas were attacked. In the absurdly polarized politics of this province, ideas are seldom, if ever, considered on merit; they were tagged as left or right, then summarily dismissed. For daring to frame commercial fishing from an economic model, Pearse was labelled a corporate facilitator, a shill. And it hurt.

While his own country may have rejected the notion of individual quotas, it has grown in currency elsewhere. By the early 1980s, several countries, notably Iceland, New Zealand, and the Netherlands, adopted or were experimenting with individual quotas. The resulting improvements were impressive. With the security of defined shares of the catch, and the ability to adjust them through purchase and sale, fishers restructured their operations to achieve economies of scale and harvest their quotas as efficiently as possible, or alternatively sold them to others. The excess capacity of fleets was soon eliminated by the fishers themselves.

In addition, now freed from the race for the fish, fishers worked for most of the year when markets were most favourable, and took time to clean and prepare their fish for the best prices. With higher prices for their product and the lower costs of rationalized fleets, fishers made more money.

Pearse's idea was catching on around the world. And, since retirement, he has become a kind of crusader, when it comes to individual quotas. During an interview, he makes his points in a series of bullets; he will not be redirected.

Under traditional fishing regimes, he continues, fishers regarded each other as competitors, a clear barrier to co-operation and information sharing. But once their catch shares were defined, fishers found they had a common interest in co-operating to protect the stocks, sustain and increase the total catch, enforce the rules of fishing, and generally to improve management—not just to protect their catch but also to increase the value of their assets in fishing rights. Today, fishers operating under individual quotas often collectively take responsibility for (and pay for) detailed monitoring of catches, biological sampling, surveys, administration of quota transfers enhancement projects, and other management functions.

Such arrangements can work here in Canada. The Pacific halibut fishery is a case in point. A couple of years ago, to resolve a conflict arising from an expanding recreational catch at the expense of commercial fishers' quotas, the Minister of Fisheries fixed the recreational sector's share of the catch at a percentage somewhat greater than its actual catch at that time, and declared that if recreational fishers ever want any more, they will have to buy quota from the commercial sector. In the meantime, the recreational fishers have been leasing their surplus allocation to the commercial sector, accumulating a substantial financial endowment they can use to acquire a greater recreational share in the future if and when they need it.

One of the biggest obstacles to the successful application of individual quotas, according to Pearse, is the concept "of the commons." It is, in fact, the great challenge of fishing around the world. As Pearse explains it, in a commons, where shares of the catch are not specified, each fisherman's economic survival is predicated on his ability to fish as hard as possible, whenever possible. In other words, to steam out to the Grand Banks or Dixon Inlet to kill as many fish as he can *before* the competition does.

Pearse then launches into an expansive soliloquy about why, in his opinion, the Grand Banks cod and so many of the planet's other fisheries are being devastated: the tragedy of the commons. The tragedy of the commons occurs when resources can be exploited by anyone. In open-access fisheries, no fisher has an incentive to leave any fish behind to breed because he knows that the next fisher that comes along will simply take whatever fish he leaves, and subsequently sell it. It's a race to the bottom with both fish and fishers losing out. This race to the bottom has all but decimated the Grand Banks.

A classic problem, he explains, and one that can only be resolved when fishermen set and manage their own quotas while governments get out of the business, except perhaps to provide the legal infrastructure. There is a proviso, though: no management regime will work in a climate of mistrust, suspicion, where poaching is, in some quarters, considered a birthright.

Pearse has won over several long-time critics. In an editorial he wrote for the *Victoria Times Colonist*, author Terry Glavin suggests that a system of individual quotas heralds an "unprecedented hope for all those imperiled salmon runs, and for the prospect of a reconfigured, sustainable fishery, along with an equitable and honourable cessation of the fishery's debilitating Indian wars." He writes:

> The key to all of this lies in a novel approach to fisheries management that's been quietly proving itself in several B.C. fisheries over the past few years. Halibut, sablefish, sea urchins, geoducks—all these fisheries are fairly sustainable. Even the big-boat trawlers, the bane of environmentalists, are in on it.
>
> I never thought I'd be advocating individual-quota fisheries publicly, so all this is a bit cathartic for me. I'd long opposed the idea, owing mainly to the sulphurous whiff of free-market idolatry that came off of it. Among many of my comrades in the environmental movement, quota fisheries are still anathema— a roadmap to corporate serfdom and the death of fishing communities. But the evidence is now in.

Glavin notes that until the early 1990s, B.C.'s halibut fishery was just like the salmon fishery. Almost all the catch went to a handful of fishing companies. The prices were lousy, the season had contracted to six days a year, and the work was dirty and dangerous. In 1991, the halibut fleet agreed to an experimental system of individual quotas; now, the season lasts all year, and the fishery is enormously profitable. Halibut stocks are healthy, and conservation concerns, such as the fleet's bycatch of troubled rockfish stocks, are being addressed in sensible ways.

Disappointing many B.C. greens, Glavin, a self-taught fisheries expert, says it will be a challenge to introduce individual quotas into the B.C. salmon fishery, because as the last of the old-style industrial "derby" fisheries, it is grotesquely subsidized, grossly overcapitalized, and it's targeting dwindling stocks, often in mixed-stock fisheries, at dangerously high harvest rates. But it can be done and should be done, he says.

Can the whole of the Grand Banks work under a system of individual quotas; will the fishermen buy in and play by the rules? This was the same province, recall, where in 2003, cod fishermen in Newfoundland and Labrador set up roadblocks on the Trans-Canada Highway, occupied federal fisheries offices, and burned a Canadian flag. They were protesting against a federal decision to shut down several cod fishing grounds on the Gulf of St. Lawrence to save what remained of stocks that had been in collapse for more than a decade. The action threatened the jobs of 1,000 fishermen and 3,000 processing workers and jeopardized the very existence of several coastal villages. Then-premier Roger Grimes announced his government would do nothing to help federal authorities prosecute fishermen who skirted the new ban.

"There's so much mistrust with the federal government," commented Sean T. Cadigan, a historian at Memorial University in St. John's, Newfoundland. "It's grounded in a lack of faith in federal capacity for fisheries management."

But, these are just mechanics, Pearse claims, convinced of their veracity. He argues trenchantly the key to modern, sustainable fisheries management is an economic one. In a sense he has married an ancient industry with modern-day business principles and practice. He contends that, to make a reasonable living for their families, fishers around the world have to stop being simply harvesters and start being managers as well. Reducing the race for the fish, he argues, means that effort can be spread out over a longer period of time, giving fishers greater control and, ultimately, their profits.

Over the years, he introduced fisheries to notions such as market economics and private property rights, anathema to people such as the former leader of the United Fishermen's and Allied Workers Union, Homer Stevens, a proud card-carrying Communist.

The extent to which fishers, responding to economic incentives, can be relied upon to allocate catches and manage their fisheries for maximum value depends critically upon their ability to control their supply of fish, which in turn depends upon the scope of their fishing rights. This link between the rights of fishers and their ability to manage is key to the successful development of market-based fisheries management regimes, insists Pearse.

For centuries, he says, fishers had no rights and no control over other fishers or potential fishers. This was appropriate as long as the supply of fish exceeded demand and fish were (or were perceived to be) inexhaustible. In these circumstances, fishers had neither the means nor the incentive to organize themselves and participate in management.

Individual quotas, on the other hand, substantially strengthen the rights of fishers and restore their control over their catches. Their right to a defined harvest eliminates the wasteful competition and interference from others. The right to transfer their rights allows them to rationalize their operations. Their proportionate interest in the catch gives them an incentive to cooperate with each other to manage their fishery and the resources they depend upon.

For these ideas, Pearse has been savagely and publicly ridiculed by union bosses, environmentalists, and coastal politicians. He was accused of being an advance man for privatization; a corporate sell-out. This criticism stung. Especially hurtful were charges claiming he had little concern for the small-time fishers who worked and lived in the many communities that dot the B.C. coast.

He was cast as a cold-hearted technocrat who sacrificed a way of life to his economic theory. Unfair, protests the man who grew up in relative poverty on what he describes as a "Vernon stump farm" with seven siblings.

As a young timber cruiser in the 1950s, he recognized first-hand that land would only grow in value. Working one summer on Lasqueti Island—nearly inaccessible in those days, it now has a pedestrian-only ferry and is known for its illegal high-grade marijuana—and noting a 200-acre parcel of land on the southern tip of the island was for sale, he borrowed the money to close the deal. Several years later he built a log home on a rock bluff with a 180-degree view of the Georgia Strait. With little infrastructure and services on the island, he built his own systems for water, sewage, and lighting.

Thus began a life-long interest in the history and economics of the Pacific fishery, and in fishing around the world. Over the years he would introduce market economics and business models.

Meanwhile outside the academy, something happened. Interest in ecology and oceans grew naturally from environmentalism. A new generation of children was visiting aquariums, watching *Free Willy,* and studying to become ocean biologists. This new awareness stands as a direct challenge to the catch-and-kill mindset of commercial fishers on both coasts.

Fishing *had* to change. In the summer of 2007, in what could be seen as a crowning glory for an idea transformed into practice, Pearse delivered the keynote speech at an international symposium on fisheries economics in Iceland. His subject was brought into tight focus, particularly in the wake of the 2006 *Science* article mentioned earlier, in which several of the world's leading marine biologists concluded that all fish and seafood species worldwide would crash by 2048.

The journal also estimated that 90 per cent of the world's species of large predatory fish are already gone; overfished, some to near extinction.

Transferable quota systems have been tried in a range of countries, including New Zealand, Australia, America, Chile, Iceland, and the Netherlands. Others, such as Argentina, are thinking about them. Where they are in operation, the fishermen have begun to use words such as "product" and "customers." They complain that the joy has gone out of the life—but they are making money.

Significantly, the countries that have led the reorganization of fisheries and have benefited most from it—notably Australia, New Zealand, and Iceland—have all adopted new legislation and administrative structures to accommodate their new regimes. But all three are isolated and more politically cohesive than Canada.

Still, both Pearse and Pauly are convinced their ideas would work on the Grand Banks. Each, in his own way, warns that real structural change will be built on new ideas and a new mindset that evolve from the fishermen themselves and not politicians and mandarins from St. John's or Ottawa. Despite differences, personal and professional, both men offer new ideas that could be used to rebuild the Grand Banks cod. Ironically, one set of ideas complements the other, because no one "fix" will work, as the stocks have been beaten to near extinction.

Brave New Ideas, such as those from Pearse and Pauly, may over time point to a new way for beleaguered fisheries. Given the near extinction of the cod stocks there, it makes biological sense to halt all fishing on the Grand Banks and turn the area into a marine-protected area for years, even decades, until the stocks regenerate. Then—and only then— could a comprehensive system of fair and equitable quotas lead to a new management regime based on enlightened self-interest, because the fishermen would, under a quota regime, be transformed into small-business people, eager to work with scientists and managers to make sure the fish returned next season. A system of quotas would reduce or eliminate the dirty little secret of outport Newfoundland: poaching.

While it is impossible to get any real data on poaching, it is widespread. Even a visitor will note its evidence. In every village and outport, cod is for sale—in restaurants, pubs and with a knock on the back door. Along the newly paved highways, parked on the on - and off-ramps a handwritten sign adorns a parked car with a middle-aged figure slumped inside: Cod for Sale. Many visitors, aware of the cod collapse, are surprised to see the fish. Waiters and servers, when queried, resort to the wink-wink of a Monty Python skit. Driving the peninsulas of the province, fiddling with the radio dial to kill the hours, a visitor notes the constant airplay of a series of provincial Fish and Wildlife public service announcements, in which a stentorian voice reminds the locals of the penalties for poaching fish and killing moose. Indeed, one reporter who insisted on anonymity told me that poaching—non-compliance in federal jargon—is considered a birthright by many Newfoundlanders. A time-honoured bit of theatre is played out to this day: fishermen and hunters chat and exchange pleasantries with the local fisheries officer. Once the officer has driven down the road, the locals climb in their boat or wander out on the heath in search of dinner. This, the reporter said, was the way of survival for five-hundred years.

Attitudes and mindset have much to do with fishing. For too much of the history of the Atlantic fisheries, the wrong people have been making the wrong decisions for the wrong reasons. Politicians have subsidized expansion of the fishery despite countless warnings of overcapacity. They have permitted catch levels far beyond those recommended by their own scientists. With an eye on the next election, they have chosen actions with short-term political pay-offs and disastrous long-term consequences.

Faced with successive crises, politicians and bureaucrats have proven to be adept at studying the problems of Canada's troubled fishing industry. Studies have provided excuses for inaction that has served the government's interests. As fisheries biologist Carl Walters noted, bureaucrats "are rewarded not for effective action, but for making every problem disappear into an endless tangle of task force meetings and reviews."

An endless tangle indeed. In the last century, well over 100 official commissions have reviewed the numbers, consulted with stakeholders, and penned volumes of recommendations that have gathered dust on governments' shelves, in what one of those countless commissions described as a "traditional cycle of a crisis, followed by a study and perhaps a subsidy, then partial recovery, then back to a crisis again."

Although governments have succeeded in putting off some fisheries problems, they have not succeeded in solving them. How could they? Too often, they themselves have been the causes of the problems. As Member of Parliament and former fisheries committee chair George Baker once said of the groundfish collapse, "This is not a natural disaster that's happened." Speculation abounds that Baker was fired for his sharp criticism of the government.

To this day, no politician or bureaucrat has been held accountable for making the decisions that destroyed the groundfish stocks. None has been fired, or demoted, or even, unlike the scientist who dared tell the truth about the collapse, reprimanded. Indeed, two key DFO science officials—Jake Rice and William Doubleday—have since been transferred to Ottawa.

In January 1998, fisheries biologist Ransom Myers commented on the collapse of the cod: "The disaster in the cod fishery is now worse than anyone expected . . . It may be a generation before we see a recovery of the cod. That a 500-year-old industry could be destroyed in fifteen years by a bureaucracy is a tragedy of epic proportions."

Fishing is about values. Unless there is a fundamental shift of values and a new way of looking at the Grand Banks, the ideas of Pearse, Pauly, and many other experts will come to nothing. Indeed, in what has become an academic industry, several disciplines now study other aspects of the Grand Banks and other fisheries including social sciences such as anthropology, sociology, and history. Many of these look back in time to document the plight of the little guy, showing in readers a Pavlovian sympathy for the plight of the lone fisherman forced into bankruptcy by the evil of market forces in an uncaring global economy. Touching and often unrelentingly gloomy, such narratives have themes much in common with the loss of the family farm on the prairies, the shuttering of the canneries on the British Columbia coast, and the death of mom-and-pop retailers when Wal-Mart sets up shop in the neighbourhood.

The Brave New Ideas of Pearse and Pauly may be moot because, today, many Newfoundland fishers are pleading for the right to power up the dragger fleet on the Grand Banks once again. Out on the Rock, far better empty seas than lost elections.

The collapse of this stock offers documented proof that politicians and senior officials were willing to sacrifice a renewable resource to protect jobs and hold onto political power. Again and again, in Newfoundland, politics trumps science.

The fishermen themselves, too often patronized and mythologized as symbols of bravery and independence—even if we suspect that some may be impractical and irresponsible—deliberately clinging to a dangerous and outdated trade whose apparent demise has probably been brought about through their own greed, sanctioned and subsidized by government subsidies.

Incredibly, there may be more to come. Against all scientific evidence, fishermen claim some of the cod have come back—in their inlet and in their bay. Federal politicians, such as federal fisheries minister Loyola Hearn whose riding is Renews, Newfoundland, are attuned to every rumour in the outports. Some suggest Hearn is making noises that it may soon be time to open more commercial cod fishing on the Grand Banks again.

Sheer lunacy. Meanwhile, out on the Grand Banks, they are dragging for shrimp, scarring and bulldozing the very spawning and rearing grounds the cod need to survive. There is an inverse ratio between the two species and, today, shrimp is far more profitable. Dr. Johanne Fischer, executive secretary of NAFO explains: "[T]he more shrimp we fish, the less is left for groundfish such as cod, a vicious cycle. So, if we want the stocks to be restored and see as many cod on the Grand Banks as Cabot did, we probably better stop fishing everything and everywhere around Newfoundland and offshore for the next couple of decades—and then start with a traditional 1700s fishery. And yes, I'm serious."

12

A Tale of Two Newfoundlands

"We all know in our hearts that Newfoundlanders deserve at least some of the blame."
—Glen Power

Fishing, in Newfoundland, until the Grand Banks collapse, was not just part of the culture—it was the culture; stamped indelibly on every aspect of political, economic, and social life. Did a world view, produced by the vicissitudes of 500 years of hardscrabble isolation, kill the Grand Banks? Did a people with a hunter-gatherer mindset destroy one of the world's most important renewable resources? If so, how did they make it a way of life, while somehow deflecting blame and turning victimization into a badge of honour? Are there plausible reasons to point the finger at Newfoundlanders themselves? The fact that this island has been predicated on one single reality: the harvest and exploitation of its natural resources of cod, seals, moose, and minerals—and today, offshore oil and gas?

The fact is that its geography and climate led to the development of an economy with few options for those who did not want to go fishing or work in the plant for minimum wage. Young people who today reject a life defined by the "six-and-two" of various government make-work programs would have to "go down the road" to factories in Ontario or oil rigs in Northern Alberta. To a visitor, Newfoundland offers a classic example of the human need, and will, to survive.

Any visitor trying to make sense of this exotic, almost foreign place will come face to face with a series of reminders that this 500-year-old culture was until recently based almost exclusively on the catching and killing of the Grand Banks cod. For the British overlords, this was a temporary colony they looked forward to leaving in late fall, for the gentler, more forgiving climates of Bristol and Southampton. Newfoundland was a brutal place, tough and cold, with its own bloody history.

To survive here in the old days required a fierce self-reliance. Today, since the death of King Cod, new attitudes have replaced the proud independence: a new sense of entitlement and victimization that many say is a direct result of the Unemployment Insurance Act introduced in the mid-1950s. These attitudes have been skillfully manipulated by politicians who, since the day the province joined Confederation, have played on the ambivalent relations with the rest of Canada. Newfoundland premiers have, over the years, played the have-not card with varying degrees of success. That card is still being played today by Premier Danny Williams.

Federal-provincial relations were one of the hot-button issues during the 2007 fall provincial election campaign. Relations between Danny Williams and Prime Minister Stephen Harper were at an all-time low. "We've had a lot of rivalries and political conflicts between us and the government of Ottawa, but this, I think, is the most extreme," said former federal cabinet minister John Crosbie, no stranger himself to strained relationships between Newfoundland and the federal government.

In fact, Mr. Crosbie has said that the current rift is worse than disagreements the province had with former Liberal prime minister Pierre Trudeau: "There were some very rough days between the former premier Brian Peckford administration and Mr. Trudeau, but this is probably the most obstreperous and publicly vindictive battle that we've seen, I think, in our fifty-eight years of being in Confederation," he claimed.

While Mr. Williams trumpeted the province's fiscal turnaround thanks to a gush of revenue from the offshore oil sector, the leader of Newfoundland and Labrador's Liberal Party, Gerry Reid, was focusing his campaign on rural issues such as out-migration, the death of rural Newfoundland, and the cod collapse.

On the ever-present talk shows on the Rock's airwaves, the chatter is incessant, rancorous, finger-pointing, and bleated by leather-lunged insiders. For a visitor, those who call in offer an introduction to a rich and evocative Newfoundland vernacular where words have hidden tribal meaning. Words such as *buckly, black, taut, tished, loose, slatchy, slob, way* and a planoply of others amaze and confuse. This is a rich Shakespearean language, thick with puns and the dryly musical accent and vernacular of working-class Newfoundland dialect, an uncertain intonation can lend a deadpan or quizzical quality to the simplest of utterances, they parse as a foreign tongue indeed to a mutual fund manager on Toronto's Bay Street or a geologist in Calgary.

There is a serious disconnect between the romantic notion of the life and culture of small-boat fishermen—beer and fiddling, acapella laments, open-tuned guitars with minor-key tunings.

Over a skim milk latte at a high-end St. John's coffee shop, forty-one-year-old Glen Power folds up his Mac Powerbook to consider a visitor's questions. Slim, dressed in black from head to toe, he represents the new urban Newfoundland and he would fit in with the high-tech sophisticates of Toronto or Seattle.

A commercial property appraiser, he is upbeat about real estate in St. John's, a city, he believes, is growing.

Like almost everyone in the province, he once had ties to the fishing industry, but no longer. Like many Newfoundlanders, while he doesn't work in the fishery, its health and sustainability is still very important to him—an integral part of the fabric of the society, helping to define who Newfoundlanders are as a people. "The loss of the northern cod was as heartbreaking to me as it was to any fisherman," Power says.

Sipping his latte, Power carefully closes down his laptop, to share more of his views on the death of the cod. At a nearby table, former *Codco* star Mary Walsh is chatting with a friend. Power watches her out of the corner of his eye. No over-the-top, outrageous characters this afternoon; she's off camera.

"First of all, I should say that I am indeed sympathetic to the plight of fishermen. It's a tough, dangerous job where you put your life on the line almost every day. The fish are not always there, so there's a lot of risk associated with maintaining a stable income. I know and admire many guys who do this for a living, but . . ." There are lots of buts.

Ever the diplomat, Power suggests that the mindset of Newfoundland fishermen today is still rooted in a world of constant danger and struggle with the elements—a world where winds and tides can toy with even the biggest boats, spelling danger and hardship in order to pay the bills.

This may explain why a new generation no longer wants to go to sea and sheds light on Power's own ambivalent attitude towards those who still go down to the sea in small ships. He can admire, and even envy, what is seen by many as freedom and adventure while, at the same time, harbouring the insistent suspicion that many are clinging to a dangerous and obsolete trade whose apparent demise has probably been brought about through their own greed.

He finishes his coffee. Would a visitor like to accompany him on a ramble along Water Street north to Signal Hill? Perhaps to help formulate his blunt, sometimes fearless appraisal of the attitudes of his countrymen and how they've contributed to the cod's demise.

"I think those of us who work and live in St. John's have drifted away from the realities of rural Newfoundland," he says as we walk along on a sunny late summer afternoon. "In essence, there are two Newfoundlands today: this one, and the rural one."

Power believes saving rural Newfoundland is *the* issue of the day and that the solution must come from rural Newfoundland itself. He also believes it will require a painful and difficult shift in attitude. "There has to be a change in the way things are approached and done. There needs to be more innovation, entrepreneurship, ingenuity, and a focus on sustainability. Maybe for a start, fishermen should take a more business-like approach to the industry—this could be a catalyst perhaps. I hear that such an approach is becoming more and more the norm these days, which is good news." The apparent lack of concern by some fishermen towards conservation and the environment is a major sore point for Power.

As we wend our way through the Battery, a community of tiny homes perched over the narrows of the Harbour, we can see, up above, the stone fort on Signal Hill, the English bastion built in 1762 to fight the French. A system of trails—locals use it for exercise, competing in summer with busloads of tourists and digital cameras—leads to the high ground of the fort. Power, by this stage, has become fearless in his commentary. He steps into the breach of the iconic: he questions the mythology of the fishery, the core of the island itself.

"Many here say the fishermen seem happy enough to decimate one species and then move on to the next. Maybe that's not the case—I don't know—but it's what many people here think. If this isn't the case—well, maybe they need to do some PR."

He tries to explain the strength of his feelings as we sit on a stone bench looking east over the Atlantic Ocean. "I used to listen to the fisheries broadcast—useful when I was doing valuation work on fish plants and marine properties. Most of what I'd hear was whining and complaining about not being able to catch more, whining about fisheries closing early, whining about conservation measures in general. There was never too much positive talk of conservation, not that I heard anyway. I just stopped tuning in—it was too negative most of the time and hard for me to listen to.

"If overfishing was the cause of the northern cod disaster, then we all know in our hearts that Newfoundlanders deserve at least some of the blame. I've never heard too many fishermen own up to this though; they usually point the finger at Ottawa and the foreign devils."

Power agrees the cod disaster may have been partially due to bad management by government, but asks who has continually exerted political pressure on government, and in turn on the system? "Check the news archives to see how many government offices have been thrashed over the years," he says. "How many protests have turned violent, and how many politicians have given in to the pressure from one group or another?

"The fishermen and rural communities have continually exerted intense pressure on the politicians to facilitate the opening of quotas, the increasing of quotas, the opening or propping up of fish plants, subsidizing of fish plants, subsidizing of income, stopping or altering conservation measures, etc. This is still the case today. Is there another industry in this country with so much political and government intervention?"

Power, representative of an emerging urban class, acknowledges the need to provide for families and maintain a way of life, but he believes the system was pushed, aggressively and mindlessly, way beyond a sustainable level. "There were fish plants in every other bay and cove—a huge industrial infrastructure along thousands of kilometres of coastline, most of which only worked for a portion of the year."

He is convinced that if every community and every plant wasn't subsidized, Newfoundlanders would have found a way to become more efficient, more profitable. Maybe fewer plants and fewer employees would have resulted in better year-round living, with fewer fish taken from the sea. "Of course to suggest such a thing would have started a political firestorm," he says. "Instead, when the moratorium was imposed, the whole house of cards came tumbling down."

On July 2, 1992, the day the cod moratorium was announced, about 1,000 construction workers were busy building a huge base structure for the Hibernia oil fields in Bull Arm, Trinity Bay. Would oil and gas help rebuild Newfoundland, turning it once again into a land of opportunity, a place for the young and enterprising, a place with a future? The answer, it seems, depends on where in the province you live. It certainly hasn't helped Power's rural Newfoundland.

Yes, the revenues are promising. In its 2007/2008 budget, for example, the provincial government forecasts more than $1.03 billion in offshore royalty revenues. That's more than double last year's oil revenues. If the forecast holds true, oil royalties will account for $1 out of every $4 the province collects in taxes, investments, and fees. A little fiscal breathing room not seen in years.

But the Avalon Peninsula—which accounts for almost half the province's population—has enjoyed the lion's share of the jobs, construction, and spending associated with developing and producing oil from three offshore fields. It has meant a residential housing and commercial building boom, increased salaries for some, and the lowest unemployment rate in recent years.

But rural communities that depended on the cod fishery haven't fared as well. They're still losing people, young and old, to urban areas on the Avalon or to other parts of Canada. At the same time, there is some mitigating good news as a fledgling tourism industry begins to emerge from, in many cases, the hard-hit fishing villages.

Mid-morning in high summer the Gatherall clan was loading another batch of tourists onto the catamaran for a ninety-minute spin out through the narrow harbour to the Witless Bay Ecological Reserve for a close-up. Visitors from across Canada, the United States, and Europe pay forty-nine dollars each for the tour in hopes of seeing the puffins, seagulls, and humpback and minke whales. Heading east leaving the harbour this specially built catamaran hit a bank of fog. Skipper Al Gatherall, the eldest of the two sons looked worried. As a twenty-three-year veteran of the tourist trade, he wants his guest to have a memorable experience that they will talk up once they get back home. To drift about in the fog, even on a day of calm Atlantic sea is bad advertising indeed.

The Gatheralls, Irish Catholics who first came to the Avalon Peninsula in the 1600s, are no strangers to adversity. For generations, the families were successful cod fishermen who became a dominant force in village life. Resourceful and brave, they hauled their catch to the local fishing plant for more than 400 years, adapting to changes in technology until the years leading to the cod collapse. Unlike most of their peers, the Gatheralls had recognized the jig was up and even before that disgraceful day when Fisheries Minister John Crosbie announced the moratorium on all cod fishing, the Gatheralls had been making their plans to get out of the dying industry and to use their skills as seafarers to ferry tourists out into the Atlantic.

Over time, they built their own twenty-metre-long catamaran, a restaurant, and gift shop with branded T-shirts, fleece jackets, and hats. To keep costs down, they converted the boat house where they built the Gaffer VI into their emporium. Inside, younger brother Mike, looks up from his computer and turns off his telephone to explain the 2007 season has, to date, been plagued by poor weather that is keeping the tourists away. His fears were later realized. In December he reported a so-so year, "on par, or slightly better than 2006."

Paralleling the harbour, a gravel road connects to the East Coast Trail. Fragrant wild pink roses and purple thistle crowd the rocky but well-maintained trail with its postcard views of Bay Bulls and steep cliffs. Near the lighthouse, several hectares of land have been recently purchased by a Toronto real estate developer, who, it seems, has plans to build a gated community for affluent visitors now increasingly discovering the province's charms at rock-bottom prices. Its proximity to the world-famous East Coast Trail, the island's low crime rate, and the natural beauty of this part of the Newfoundland coast will very likely be featured in the marketing brochures to attract potential U.S. and European investors.

The proposal has shaken and divided the local council. Bay Bulls is being transformed. Once a tight-knit community of Irish Catholics, it is now becoming a bedroom suburb of St. John's, just twenty minutes to the north. This transformation caused some serious soul-searching as the issue is sporadically played out in community newspapers.

Taking our leave, we walked past the Holy Trinity Anglican cemetery, waving to an old man bent over the family plot. He was using Roundup to control the weeds. We stopped to talk. And once he confirmed our visitor status—this is accomplished by simply opening our mouths; Newfoundlanders use verbal delineations in the same precise way the English can tell class and rank in a few short sentences—he anticipated our questions. "The cod stocks collapsed from overfishing," he said, the consequence of ignorance and greed. (In what proved to be a very frank assessment, what he described as ignorance was a direct result of living on an island, being separated from the ideas and knowledge of the continent. "If I ever had any say in my reincarnation, I'd be born in Nova Scotia. Sometimes it is impossible to get off this island—and expensive, too.")

Power grows quiet. Time for a visitor to take his leave? But wait. One more thing the visitor must hear, must understand: His impatience, his irritation, his fury at the stereotype of lazy Newfie, the surly ingrate who has polished the politics of victimization to a high art. He tells of attending conventions in other Canadian and U.S. cities and waiting for the inevitable: after dinner and drinks one of his colleagues will launch into a Newfie joke—or two. Over the years, Power has grown fiercely resentful of these boozy jibes, no matter how well intentioned. A proud professional, he seethes under the yoke of the churls.

He has much to seethe about. In 2005, *Globe and Mail* columnist Margaret Wente, after first declaring her great admiration for citizens of the Rock, described some of its citizens as "surly" ingrates, "gobbling" cod tongues while they luxuriate in a great "scenic welfare ghetto." In general, she put down everyone in Newfoundland as part of a set of lazy, self-indulgent, whining spongers:

> No one is better at this blame game than the Newfs, egged on by generations of politicians. The only way to get elected there is to pledge to stop the terrible atrocities of Ottawa (i.e., not sending enough money). If you should make the error of suggesting that people might have to become more self-sufficient, your political career is dead. Politicians like to get elected, which is why things never change.

Favourite Newfoundland son and intellectual star Rex Murphy, also a columnist for the *Globe*, fired back with a fierce retort. The snit, which occurred just as blogging took off, ricocheted around Newfoundland for years, with many taking sides and issues. Murphy slammed Wente for her stereotypes, pointing out "I have known people so hostile to every notion of something for nothing, they wouldn't trouble a neighbour to borrow a cup of milk. I've known legions of men and women who put in a lifetime's work of a kind that those of us who spray words for a living should be embarrassed to stand next to."

But when it comes to the collapse of the cod fishery, Murphy evinces a curious obtuseness:

> Then there's this business where Margaret writes of Newfoundlanders blaming us for the collapse of the fishery. Who's this "us"? The citizens of Canada didn't collapse the fishery, and no one in Newfoundland even dreams they did. The only point

on which any blame is being assigned is over the stewardship of the resource since Confederation. That was federal. No one argues otherwise. And it is surely fair, and not victimhood, that if the government that had control failed in its stewardship, then it should bear some responsibility for so failing.

It has to be said: Murphy, presumably representing the views of many Newfoundlanders, appears to be abdicating responsibility for the Grand Banks cod collapse. Reverting, in time-honoured island fashion: blame the other, blame Ottawa.

But Ottawa did not plunder the Grand Banks to near extinction. It produced botched science and weak management, yes. Seals did not slaughter the cod. Nor did hot, nor cold water. Nor inshore fishers. Nor the foreign devils.

New data, announced in these pages, proves conclusively the decimation falls squarely on the Canadian offshore dragger fleet, sanctioned, legitimized, and subsidized by politicians—and sent out to sea by St. John's and Ottawa.

Still everyone, had his and her pet theory. In no particular order they included: seals, foreign devils, cold water, hot water, climate change, bad science, bad management, federal politicians, fisheries officials, and the Canadian offshore fleet.

Several wondered whether the question of who killed the Grand Banks would result in "the good German" defence: here on the Rock people were just doing their job, surviving, paying the bills, going along to get along. But this slide into moral relativism leaves the mystery unsolved. And the collapse of the Grand Banks cod is far too important to resort to mewling new-age forgiveness and apologetica.

It is critically important to assign blame, to take responsibility so the people of Newfoundland can take a corrective course of action.

While Newfoundland was the first to destroy a rich and productive cod fishery that once fed the world, it likely will not be the last. Indeed, new lessons to be learned—not simply biological or economic fixes— must be built on new ideas that resonate in a changing island culture.

Fishing, like so many industries, must be stripped of its romance, charm, and mythology. Fish equals food, and food equals money, and that is why men brave terrifying seas to catch and kill them.

Many Newfoundlanders still live by the myth of the small-boat fishermen, bounding bravely out of Dildo or Petty Harbour. Things are unlikely to change until this iconography is relegated to the heart-wrenching songs and poems of history. There is a major difference here: it is right, noble to respect the rich history and traditions of the past. But quite another to have it ossified in amber so that new ideas cannot penetrate.

It is a fact of Newfoundland life: Outport life is dying, the plants are closed, the fleet idles at the dock, young people are leaving in biblical numbers, and once-thriving communities are transformed into a dispiriting array of T-shirt shops and souvenir stands harking back to another time.

The Canadian offshore dragger fleet killed the Grand Banks cod. The fleet will—ever manipulated by local politicians forever playing the victim card—move on to the next, decimating lobster, crab, shrimp, and other stocks.

Packing up to leave Newfoundland and return to the West Coast, one cannot help but think of a troubling view of the future, one perhaps without fish. In a taxi on the way to the newly refurbished St. John's airport, I wondered whether the Grand Banks cod collapse is just the first of many. Whether Pacific salmon, tuna, and swordfish will soon follow in other parts of Canada, other parts of the world.

Recent figures show more than 70 per cent of the world's commercial fish stocks are exploited, overfished, or collapsed as result of advances in technology, increasing demand for product, and government subsidies. Meanwhile, a few success stories of increased catches after establishment of no-fish reserves and regional shutdowns exemplify resiliency of fish when habitat is not severely damaged. At the same time, a growing public awareness about fisheries and oceans, skillfully articulated by conservation groups who use blogs and the Internet to connect growing numbers of students and young people, propelled by the ideas of Al Gore and David Suzuki, demands change on the Grand

Banks. This new energy—and above all, a new sense of optimism—is sorely needed in Newfoundland's fishing industry.

On October 18, 2007, Danny Williams swept to power because of the public perception he successfully forced oil companies to capitulate to demands for a greater share of their energy windfall. The premier is rightly seen by the people of this island as a fighter, a dynamic politician determined to employ the new ideas needed to transform the province into a modern economy. When it comes to renewing the cod stocks of the Grand Banks, is he willing to consider new ideas such as marine-protected areas? Will his enthusiasms wane, as he, and his constituents, resort to the time-honoured game played out here on the Rock: blame Ottawa for everything? Take no responsibility. Rack up votes at the polls.

There is always reason for hope. Ironically, it just might come from a new generation of Newfoundlanders, the young people now finishing high school and moving on to post-secondary studies in unprecedented numbers. In the past, the young men would have almost certainly joined their fathers, fishing out on the Grand Banks. Today, they are taking courses— biology, oceanography, chemistry, and offshore engineering— that will provide insight into the Grand Banks collapse and crisis. Glen Power's words echo in my head: "There has to be a change in the way things are approached and done. There needs to be more innovation, entrepreneurship, ingenuity, and a focus on sustainability."

The shift in attitude. Fundamental. Defining. The Future, as people learn to appreciate and protect the seas, to see a world beyond the foggy shore. Awareness is the root of the solution and, in this rich nation, there is much demanding that we be aware.

Appendix

The Two-Hundred-Mile Fishing Limit

May 10, 1994

Canada tables legislation which will give it authority to stop vessels suspected of fishing illegally beyond its 200-mile limit and to arrest flag-of-convenience and stateless vessels fishing for endangered fish stocks. It will allow Canada to identify the straddling stocks in danger, enact conservation regulations and compile a list of vessels against which these regulations will be enforced.

SOURCE: Coastal Fisheries Protection Act, amended May 1994. Section 5.1, paragraphs (c) and (d). From http://laws.justice.gc.ca/en/C-33/35158.html.

On May 10, 1994, the House of Commons amended the Coastal Fisheries Protection Act (CFPA) in several areas. The amendments drew attention to an "urgent need for all fishing vessels to comply in both Canadian fisheries waters and the NAFO Regulatory Area with sound conservation and management measures for those stocks and that "some foreign fishing vessels continue to fish for those stocks in the NAFO Regulatory Area in a manner that undermines the effectiveness of sound conservation and management measures." The new amendments were to enable the government to take action to prevent further stock depletion while still seeking international resolution to illegal foreign fishing. To that end, the amended Act declared that:

> No person, being aboard a foreign fishing vessel of a prescribed class, shall, in the NAFO Regulatory Area, fish or prepare to fish for a straddling stock in contravention of any of the prescribed conservation and management measures.

Further, the Department of Fisheries and Oceans was entitled to enforce this section of the CFPA through protection officers:

A protection officer may

a) for the purpose of ensuring compliance with this Act and the regulations, board and inspect any fishing vessel found within Canadian fisheries waters or the NAFO Regulatory Area; and

b) with a warrant issued under section 7.1, search any fishing vessel found within Canadian fisheries waters or the NAFO Regulatory Area and its cargo.

These changes to fisheries policy can be interpreted in two ways. One, the Canadian government took it upon itself to ensure that NAFO regulations were being followed in areas adjacent to Canada's EEZ. As previously mentioned, the EU's continued use of the objection principle rendered NAFO impotent to stop over-fishing. Since EU would do nothing, it was up to Canada to take the moral high ground. The other interpretation is that Canada was once again taking unilateral steps, to NAFO's detriment, in order to extend Canada's control over the entire Grand Banks area. It was an action similar to 1977, when Canada extended its EEZ to the 200-mile limit. Canada was more interested in extending sovereignty for economic reasons than NAFO regulation and environmental preservation.

Either way, the amendments to the CFPA effectively extended Canadian jurisdiction to include NAFO waters. This change was highly controversial: both the EU and the United States expressed their concern. The EU felt that the CFPA gave too much latitude to the

Governor in Council in terms of changing which classes of ships were subject to regulation. The United States expressed its concern about the potential dangers posed to American fishing vessels in NAFO waters.

The same day that the CFPA was passed in Parliament, Canada took another step towards protecting the actions of the DFO. The government deposited an amended declaration of acceptance of the compulsory jurisdiction of the International Court of Justice (ICJ). Like other international institutions, the ICJ was founded on voluntary membership, and nations could make reservations about certain types of disputes they did not want to fall under ICJ jurisdiction. Canada had previously exercised that right in 1985, with regard to disputes that fell strictly within Canada or in the Commonwealth, or could be settled by other means. The new reservation dealt with NAFO:

... the Government of Canada accepts . . . the jurisdiction of the International Court of Justice . . . subsequent to this declaration, other than

... d) disputes arising out of or concerning conservation and management measures taken by Canada with respect to vessels fishing in the NAFO Regulatory Area, as defined in the Convention on Future Multilateral Co-operation in the Northwest Atlantic Fisheries, 1978, and the enforcement of such measures.

End Notes

Chapter 3

1. Finlayson, Alan Christopher. 1994. *Fishing for Truth: A Sociological Analysis of Northern Cod Stock Assessments from 1977 to 1990.* St. John's, Newfoundland: Memorial University of Newfoundland, Institute of Social and Economic Research, Social and Economic Studies No. 52. 138–9.
2. Northern Cod Review Panel. 1990. Independent Review of the State of the Northern Cod Stock: submitted by Dr. Leslie Harris. Ottawa Dept. of Fisheries and Oceans.
3. Finlayson, Alan Christopher. 1994. *Fishing for Truth: A Sociological Analysis of Northern Cod Stock Assessments from 1977 to 1990.* St. John's, Newfoundland: Memorial University of Newfoundland, Institute of Social and Economic Research, Social and Economic Studies No. 52. 150–51.

4. Finlayson, Alan Christopher. 1994. *Fishing for Truth: A Sociological Analysis of Northern Cod Stock Assessments from 1977 to 1990.* St. John's, Newfoundland: Memorial University of Newfoundland, Institute of Social and Economic Research, Social and Economic Studies No. 52. 115.

5. Canada. *Treasury Board Manual: Government Communications Policy.*

6. SCOFO (Standing Committee on Fisheries and Oceans). 1997a. Transcript of Evidence, 1998.

7. SCOFO (Standing Committee on Fisheries and Oceans). Transcript of Evidence1997a, 945–50.

8. Pilkey, H. and Pilkey-Jarvis, L. *2007. Useless Arithmetic: Why Environmental Scientists Can't Predict the Future.* Columbia University Press, 2007.

9. University of British Columbia, Professor Carl Walters email to author. 2007.

Chapter 5

1. Auditor General. 1997. *Report of the Auditor General of Canada to the House of Commons: Atlantic Groundfish Fisheries.* Ottawa: Minister of Public Works and Government Services Canada.

2. Canada. 1976. Department of the Environment; Fisheries and Marine Service. *Policy for Canada's Commercial Fisheries.*

Index